AN ANALYSIS OF COLOR CHANGES
AND SOCIAL BEHAVIOR OF
TILAPIA MOSSAMBICA

BY

E. H. NEIL

UNIVERSITY OF CALIFORNIA PRESS
BERKELEY AND LOS ANGELES
1964

University of California Publications in Zoölogy

Advisory Editors: G. A. Bartholomew, Demorest Davenport, C.R. Goldman,

B. H. Levedahl, P. R. Marler, R. I. Smith

Volume 75, No. 1, pp. 1–58, 5 plates, 6 figures in text

Approved for publication November 22, 1963

Issued October 2, 1964

Price, $1.25

University of California Press

Berkeley and Los Angeles

California

◆

Cambridge University Press

London, England

PRINTED IN THE UNITED STATES OF AMERICA

CONTENTS

Introduction	1
Acknowledgments	1
Materials and Methods	1
Definition of Terms	2
Preliminary Description of Behavior	5
External Coloration	6
Schooling Behavior and Color Changes	8
Territorial Behavior and Color Changes in the Male	10
Sex Discrimination as Related to Male Coloration and Behavior	19
Courtship Behavior and Color Changes	20
Prespawning and Spawning Behavior	28
Brooding Female Behavior and Color Changes	31
Fighting Behavior and Color Changes	32
Discussion	37
Summary	42
Literature Cited	45
Plates	47

AN ANALYSIS OF COLOR CHANGES AND SOCIAL BEHAVIOR OF TILAPIA MOSSAMBICA

BY

E. H. NEIL

INTRODUCTION

Tilapia mossambica is a member of the family Cichlidae, which has long been known for the ability to change color. Baerends and Baerends van Roon (1950) and Seitz (1948), working with this species in the laboratory, described many of its color patterns in detail and noted the remarkable speed with which the fish could change color.

Until a few years ago only the most striking color patterns were used as indicators of behavioral states: e.g., the red belly of the three-spined stickleback (Tinbergen and Van Iersal, 1947). Quantitative analysis of rapid color changes in fish has been largely ignored. This unique phenomenon is especially valuable in studies of many fishes. The purpose of the present paper is to demonstrate how quantitative correlations may be used in predicting the behavior and color patterns of a cichlid fish. *T. mossambica* was chosen for study not only because of its behavioral and color changes but also because the changes in intensity of performance may be categorized and correlated. This study uses all three methods with respect to individuals and groups of *Tilapia*.

ACKNOWLEDGMENTS

For his guidance in the preparation of this material, I am greatly indebted to Peter R. Marler, of the University of California. I express my thanks to George Barlow, of the University of Illinois, for meticulously reading the manuscript and providing many valuable suggestions. I am grateful for the assistance of Emily Reid and Janet Smith, of the University of California, for their excellent drawings. My thanks to Earl S. Herald, of the California Academy of Sciences, and H. S. Bern, of the University of California, for providing the animals necessary for this study. This research was supported in part by a grant to Peter Marler from the National Science Foundation.

MATERIALS AND METHODS

The natural range of *T. mossambica* is the east coast of Africa from North Kenya to South Tanganyika. This species populates the quieter waters of small lakes, overflows, and swamps associated with the coastal drainage systems of this area (Copley, 1958). It has been introduced and extensively cultured in ponds, rice fields, and brackish lagoons throughout Southeast Asia and Polynesia where protein supplies are low (Chimits, 1955). As a result of its popularity as a food fish, *T. mossambica* is the major subject of many studies other than those on its behavior. The present paper is confined to laboratory observations, although a subsequent article will be based on field observations.

More than 1,000 *Tilapia* were employed in this study. Part of the original stock was obtained from the Steinhart Aquarium in San Francisco, and the rest was shipped from Hawaii. The fish were maintained in tanks ranging from 5 to 100 gallons. Additional observations were made on fish in a 1,057-gallon tank at the Steinhart Aquarium and in a pond 30 by 15 by 2 feet in the courtyard of the Life Sciences Building at the University of California. The temperature in the tanks was held between 26° and 29°C. The young were fed newly hatched brine shrimp (*Artemia*) and *Daphnia*. Adults received a diet of *Artemia, Gambusia affinis*, horsemeat, and alfalfa pellets. The aquaria contained two inches of white sand and from one to four garden pots split lengthwise. All vegetation placed in the tanks was eaten.

Animals were reared both in groups and singly. Because initial experiments with isolated eggs and newly hatched young were unsuccessful, owing to high mortality, young were isolated for the first time one week after leaving their mother's mouth. Swimming and feeding were the only behavioral states observed in the very young animals. Each tiny fish was placed in a separate 5-gallon aquarium which was surrounded by a partition. The fish were able to see their own reflections on the sides of the tank, but were never observed to perform courtship or aggressive movements toward them. When the animals became sexually mature at about three months of age, they were observed for two weeks to check for darkening of color and reproductive behavior. Mouthdigging was the only reproductive activity seen in males. The females were removed and placed together in a 30-gallon stock tank for later use. Seventy-four males were individually isolated. References to these isolate males are specifically denoted; all other references to male *Tilapia* are from observations on fish raised in groups.

The observations were spoken into a tape recorder and subsequently transcribed in columns indicating activity per one-second intervals ticked off on a stopwatch.

Tilapia mossambica has no common name referring to it alone. In this paper the name *Tilapia* refers only to *Tilapia mossambica*. Baerends and Baerends van Roon (1950) and Seitz (1948) described their animals as *Tilapia natalensis*. Baerends' fish were later sent to Trewavas of the British Museum, who identified them as *T. mossambica*, since the species *T. natalensis* had been synonymized with the former. Therefore Seitz' term, the "natal perch (Natalbarsch)," refers also to *T. mossambica*. Wickler (1960) notes that *T. heudeloti* of Seitz is in reality *T. mossambica*. The ethology of *T. heudeloti* has not been investigated in the laboratory.

The term "resident male" refers only to black males which have territories. All fish measurements are standard length. Terms which are not defined but were coined by the author are enclosed in quotation marks.

DEFINITION OF TERMS

Baerends and Baerends van Roon (1950) described the behavior of *T. mossambica* in great detail. Seitz (1948) placed his emphasis on the behavior of large adults. In general, I have adopted Baerends' terminology where it concerns the social behavior of *Tilapia*. Several behaviors have been added to clarify detailed action sequences; these are followed by an asterisk to avoid confusion between the two terminologies. The names of some of these new terms are not my own. Furthermore, certain behavioral components—biting, butting, and chasing—are used in the text but are not described in this section.

NONREPRODUCTIVE SOCIAL BEHAVIOR

Jerking*.—When a fish is placed in a new environment or new social conditions, it raises its median fins and uses its pectorals and caudal fin to propel itself forward. A series of short jerking movements bring the animal toward a hiding place. Baerends and Baerends van Roon described a movement which they called "jerking." This possibly refers to twiching*, which I have described as a courting movement.

REPRODUCTIVE BEHAVIOR RELATED TO FIGHTING AND COURTING

Mouthdigging.—The fish swims downward toward the sand and scoops it up in its mouth as the body is pushed forward by beats of the caudal fin. It then closes its mouth, backs up, swims a short distance forward from the area, and spits out the material. When digging occurs repeatedly in a localized area, the result is a large pit or nest. When the hole is wide enough to admit the animal's whole body, vertical digging gives way to horizontal digging in which the animal approaches parallel to the substrate.

Mouthdigging occurs in both sexes. Females have been observed to dig in the male's pit for the early stages of courtship until spawning. Horizontal digging was rarely seen with females. Males dig on assuming a territory, and continue to mouthdig at intervals as long as they remain in the territory.

Rising*.—The fish hangs at an angle of 20° with its head upward. Males may keep their median fins lowered or raised according to the circumstance. The female's median fins are usually lowered. Rising has been observed most frequently when another fish attacks. If attacks become severe, the fins are raised. I have been able to manipulate the fish in such a way that it will show rising. Baerends' term for rising is "attitude of inferiority." (It implies more about the motivational state of the animal than I can illustrate at this time.)

Intraterritorial fighting.—This type of fighting occurs between males or between females, but never between the sexes. In male combats more than two individuals may engage in the same fight. The behavioral components are as follows:

Lateral display: The body is held lateral to the opponent. All fins are erected with the exception of the pectorals, which are used to maintain balance. The branchiostegal membrane is distended.

Tailbeating: The tail is beaten sideways toward the opponent while the body is in the lateral display position. This may occur when the male is swimming forward or when it is carouseling with its opponent.

Carouseling*: Each animal circles the other in the lateral display posture, appearing to maneuver into a position slightly above and lateral to the opponent; thus both travel upward as they circle. Seitz (1948) refers to this behavior as *Kreisschwimmen*.

Frontal display: The animal faces its opponent in the same posture described for lateral display. The mouth is usually open. This is usually followed by mouthfighting.

Mouthfighting: The mouth is held open as the fish rush at each other. Each fish then tries to grip the jaws of the other. Having seized each other, they push and pull until, after a few seconds, one or both let go.

Mouthfighting, tailbeating, and chasing require a specific orientation of both animals, whereas butting and biting can occur from most postures in a fight.

Interterritorial fighting.—This is confined to nest-holding males. Its behavioral components include frontal display and mouthfighting.

Pendeling*: The fish rush at each other with the unpaired fins against the body. The mouth may be open or closed. The caudal fin brings the animal forward, and just before contact with the opponent the pectorals brake to keep the two fish from colliding. The median fins are lowered. The rival males then back up and start forward again. During the reversed swimming the median fins are slightly raised. Pendeling always occurs a number of times in quick succession. Baerends' term for pendeling was "reversing." I have adopted the more descriptive term "pendeling" from the German *Pendelin* (priv. com. Barlow).

Courting.—The components are typical for the male in the male-female courtship. The behavior of the female is described in the last courtship stages.

Tilting*: If the male is in midwater, his body is held head downward at an angle of about 30° with the horizontal. All unpaired fins and the pectoral on the side toward the female are laid against the body. The male constantly jockeys his position to assume a lateral orientation ahead of the female. If the female is in the nest, he approaches it in this fashion. Baerends called tilting "inviting." The term was changed because of its functional connotation.

Twitching*: The male maneuvers to a lateral orientation in front of the female. His body is held horizontally with all median fins folded, the pectoral fins acting as stabilizers. The pelvics are repeatedly flicked against the genital papilla. As far as I can determine, twitching has been described by both Seitz and Baerends. Baerends uses the term "jerking"; Seitz, the term "head jerking" (*Kopfzucken*).

Leading: The leading posture is similar to that described for tilting males. The caudal and pectorals are used to propel the fish down to the nest.

Twitch-leading*: The male leads down to the nest while flicking his genital papilla with the pelvics. The pectorals and caudal propel him downward, the pelvics are flicked, and the caudal and pectorals are used again. The movement downward is a series of jerking motions. The dorsal and anal fins are closed.

Rolling*: If the female is close to the nest on or near the substrate, the male circles around the nest, leaning sideways from her. The median fins and the pectoral toward the female are held close to the body. If the female is at the edge of the nest, the male rolls over on his side by means of quick beats of the caudal and one pectoral, and circles the nest in this manner. Rolling may have been what Seitz (1948) described as *Unterkriechen*.

Tailwagging: The male swims in front of and above the female and violently wags his caudal fin in front of her. The median fins are held flush against the body. Seitz' term for this movement was *Schwanzwedeln*.

Pit circling (prespawning) and spawning: "The male and female circle the nest, the male behind. Suddenly a number of eggs are ejected from her ovipositor. The female stops and picks up the eggs in her mouth. The male vibrates while ejecting sperm and the female approaches his genital pore snapping up milt and water. This process continues until the spawned out female flees the area" (Baerends and Baerends van Roon, 1950). During circling, before spawning takes place, the male and female may go through a series of vibrations and mutual nippings. At intervals the male leaves the nest to chase other fish away or to twitch and tailwag over the female. He returns in the leading position and resumes circling. Pit circling may be interrupted from time to time when the female starts to dig in the nest. The male then engages in various courtship activities until the female begins circling again. Males also pick up eggs laid by the females, but eat them. Circling was described by Seitz as *Umkehrschwimmen*.

PRELIMINARY DESCRIPTION OF BEHAVIOR

After a successful spawning, the female collects her eggs in her mouth and flees into the school. For 14 days the eggs are held in the buccal cavity and churned around for aeration and protection against fungus (Aronson, 1949). The female does not feed during this period and is extremely pugnacious, attacking fish much larger than herself (pl. 5, c). After 11 to 14 days the young are ejected in a dense swarm and begin to feed on microorganisms (pl. 5, b). They form a tight school around the mother and she takes them back into her mouth when alarmed. In the laboratory, after about three weeks the bond between mother and young slowly breaks down. The offspring wander away in groups as their locomotion improves and more food is needed. The mother takes up her young less frequently when strange stimuli are presented. If left in a small tank with her own young for a month or longer, the female may eat many of them.

Under crowded conditions, with more than thirty fish to a 30-gallon tank, the young may become reproductively mature when they are 3 to 3½ cm long. Their first reproductive behavioral acts are lateral display, tilting, biting, and chasing. These are executed in midwater. About two weeks later they descend to the bottom of the tank and begin to set up territories. When the young become sexually mature, and if numerous fish are in a single tank, four groups of activities may be observed simultaneously:

1. Schools containing both sexes are limited to certain size groups in the laboratory because the larger animals eat smaller members of their own species. These schools are in midwater.

2. Females brood eggs in the school, in the corners of the tank, or in garden pots.

3. The bottom of the aquarium is covered by nest areas, each containing a black male. These males are occupied in digging pits, defending their nests, or courting females (pl. 4, a and b).

4. Within the school, ripe females or males occasionally break away from the group and enter the territorial area. A female may suddenly follow a courting male, swim into his nest, and eventually spawn (pl. 3, b). The dark-colored schooling males may court females in midwater or lead them down to nonexistent nests, fight with one another, school, or attempt to take over the nest of a territory-holding male (pl. 3, d). The small males are usually unsuccessful in establishing their own territories, but repeatedly try to enter occupied nests.

There is no defined breeding season. A male starts digging a nest and courting whenever he becomes sexually mature. The marking of individual males revealed that, under laboratory conditions at least, male courtship is cyclic. The ownership of each territory changes about once a week. The prior residents may be found schooling in midwater and in nonreproductive coloration. These males later come down and recapture a nest or move into a vacant one. When fish are placed into a new tank, one fish starts reproductive activities and other males follow. The number of nest-diggers gradually increases for three or four days and then declines to a state of equilibrium in which the same number of animals is on the bottom for weeks at a time.

Territorial males remain in their nests at night, retaining their courtship coloration. All other fish become dark gray and most of them hover in midwater. Frequently females and occasionally males rest in the territories but never in the nests.

EXTERNAL COLORATION

Tilapia has three main types of chromatophores: melanophores, liphophores, and guanophores. Liphophores are subdivided into erythrophores (red-pigmented) and xanthophores (yellow- and orange-pigmented). Erythrophores are found along the edges of the caudal and dorsal fins of both sexes and on the pectoral fin and occipital region of the head of the male. They are most prominent when the male is in reproductive color. Xanthophores may be observed on the opercle of both sexes. These pigments disperse rather slowly. Liphophores may be partially or completely masked by melanophores or guanophores.

Guanophores occur on the body and head of *Tilapia* in more localized areas than do melanophores, and are often in very high concentrations. They may combine with melanophores to form melanoiridiophores, which give a silvery gray coloration to the fish. When alone, guanophores cause absorption or reflection of light according to the structural layering of the guanine crystals which they contain (Brown, 1957).

Fixed dendritic melanophores are densely scattered in the skin, the iris of the eye, and the fins. Within them lie mobile granules which can disperse with extreme rapidity to produce a dark coloration (priv. com. Bern). In *T. mossambica* the minimum time to produce dispersion of some melanophores was less than one second. This is confirmed by Seitz (1948). Other melanophore granules require as long as two days for complete dispersion.

Seitz (1948) found that as both males and females grow older they become increasingly dark and lose the ability to lighten to their original youthful coloration. I also have observed this inability to fade in animals that were longer than 20 cm. Therefore in most of my research I used young animals up to 15 cm long. In these young fish the melanophores are easily seen and can expand and contract rapidly. These were used as indicators of the color patterns.

Neutral Coloration

When schools of *Tilapia* are swimming, resting, or feeding, they tend to assume the color of the aquarium substrate. Over a sandy bottom the fish appear tan owing to melanophore contraction which exposes the guanophores. There are no melanistic markings except for a small black mark on the opercle, and, in the young, a round spot on the soft dorsal (pl. 4, *d*). This dorsal spot, or "*Tilapia* mark," begins to fade at about the time the fish reach reproductive maturity, and eventually it disappears (pl. 3, *a*).

Tilapia may be adapted to many different backgrounds. A white tank produces a cream-colored fish. The slightly brownish tinge is due to the contracted melanophores and yellow xanthophores. Likewise, a green fish may be obtained on a green substrate: the melanophores become partially expanded and in the opercular region the xanthophores are superimposed on them to produce the green effect. The green color is seen also where guanophores are concentrated. In a black tank the fish rapidly turn gray. A fish 6 cm long becomes completely black within one or two hours. A dull black color is produced, different from the velvety appearance of courting males. If the animals are reintroduced into the sand-bottomed tank after one day in the black tank, they fade to yellowish in about two hours. Perhaps this is an illustration of the neurohumoral action described by Parker (1948).

STRIPED COLORATION

If an animal has been startled by a strange stimulus or is chased by another fish, the Neutral coloration[1] is partially masked, either by two horizontal dark bars or by about seven vertical ones, depending on the size of the individual. These are called the Barred (pl. 5, *a*) or the Striped (pl. 5, *b*) color patterns. If a fish is handled by the experimenter, or is continually chased by another fish, a Hatched pattern appears as a combination of horizontal and vertical bars (pl. 5, *c*). In tanks with dark substrates these stripes are masked by the dark color of the fish. Over light sand a fish exhibiting this color pattern will "freeze" and finally dash for cover. The Barred and Striped colorations may be assumed within one or two seconds after appropriate stimulation, and can disappear in a very short time, but sometimes they remain even when the fish begins to assume a courting color. If courting continues for more than five minutes, the striping disappears.

The adoption of these color patterns appears to be correlated with the origin of *Tilapia*. In 1963 I observed some *Tilapia mossambica* sent from Germany to the University of Wisconsin and the University of Illinois (priv. com. A. D. Hasler and W. C. Childers). These animals had from 9 to 11 vertical bars and even those animals of over 25 cm in length were able to become Barred or Striped. The *Tilapia* at the California Academy of Sciences and at the University of California were originally imported from Hawaii. These animals have from 6 to 8 bars until they are about 4 cm long. At 4 cm the fish begin to become Striped. In fish over 7 cm long, the Barred pattern is not seen. Both the German and Hawaiian stocks exhibit the Hatched color pattern although the number of vertical stripes are different.

COURTING AND FIGHTING COLORATION

Introduction of a female into a tank containing a male causes him to pass through three recognizable color phases before reaching the fourth or reproductive color pattern. The first pattern, called Dark 1, may be assumed within one second after the male's first response to the female. The iris of the eye and the pelvic fins turn

---

[1] The first letter of each color pattern described in this section is capitalized to avoid confusion with other statements about coloration.

black and the body begins to darken in the region above the pelvics (pl. 3, *b*). As courting continues the Dark 2 stage is reached. The characteristics of this color phase are the appearance of scattered groups of expanded melanophores on the belly and caudal fin, and a black anal fin. More and more melanophores expand on the body until it appears spotted. Between these dark areas, guanophores appear as white or bluish spots. The dorsal and caudal fins become reticulated. The opercular region and the area below the eye lighten (pl. 3, *c*). These two stages are accompanied by courting movements. The Dark 3 color is characterized by black dorsal and caudal fins. The rest of the body is dotted with gleaming guanophores reflecting most of the light and situated between groups of dark melanophores (pl. 4, *c*). If a female is present, Dark 1, Dark 2, and Dark 3 are passed through quickly.

The reproductive coloration (Black) is not assumed for a day or two. Meanwhile the male passes from Dark 3 to a black coloration with three or four bluish vertical bands. This color pattern was named Banded Black. When the male becomes Black, the body is velvety in texture, the lower jaw and opercle are white, the head is brown, and the pectoral and tips of the dorsal and caudal are bright red (pl. 4, *a*). Some males had a black lower jaw and opercles from injuries incurred in digging nests or in fighting.

Once a Black male begins schooling, he may take one to two days to reach Neutral again. Some of the animals longer than 10 cm failed to become very light again once they had assumed Black coloration for more than a week (pl. 4, *a*). Once becoming Black, males never again assume the sandy-colored juvenile Neutral (cf. pls. 3, *a*, 3, *b*, and 4, *d*). Black was not observed until the fish reached a length of 4.7 cm. Smaller fish courted in Banded Black, and therefore in some of the data Black and Banded Black are lumped together under the term "Black coloration."

A female that spawns with a male assumes a dark gray body and dark pelvics and eyes. In this color she may fight with other females.

The color markers used to determine these color patterns appear or disappear quickly. Dark 1, Dark 2, and Dark 3 are artificial categories determined by the color markers. In review, the marker for Dark 1 is the black pelvics and pelvic region (pl. 1, *b*); the marker for Dark 2 is the black anal (pl. 1, *c*); and the marker for Dark 3 is the black dorsal (pl. 1, *d*). The other color patterns are easily identifiable.

SCHOOLING BEHAVIOR AND COLOR CHANGES

There are many definitions of a fish school. Breder (1951) uses the term in the strict sense to apply only to a species "... in which all individuals are oriented in a common direction, regularly spaced, and moving at uniform speed." Keenleyside (1955) suggests that the fish school is "an aggregation formed when one fish reacts to others by remaining near them." Most authors feel that there is a definite mutual attraction between individuals within a school.

Tilapia mossambica is a highly gregarious animal, surrounded by members of its species throughout life. Webster's Dictionary defines a social animal as one that "lives together with the members of its species and breeds in more or less organized communities." This definition applies to *Tilapia* both while schooling and in the territorial phase. Nest areas are cluttered with pits; so the males are in constant contact with one another. During the schooling phase, animals leave the school for only brief periods and always swim rapidly back to it. Female brooders remain with the school until their young are ready to be released from their mouths. The females then space themselves as far as possible from the nearest neighbor, but are surrounded by their young. As soon as the last young has departed, the females return to the school.

Tilapia formed tight schools from the moment they left their mother's mouth (pl. 5, *b*). Shaw (1960) was able to raise four *Mendidia* isolated from the time their optic buds were appearing. When these were introduced into the school they joined it, but were unable to maintain their position for about four hours. She notes that similarly isolated *Aquidens latifrons* aggregate in groups of two or three rather than schooling with their group-raised siblings. *T. mossambica*, although more closely related to *A. latifrons*, resemble *Mendidia* in their behavior. Any isolated young joined another animal or a school as soon as it was inserted in the tank. It then participated in the same activities as the school: swimming, nipping at the water's surface, or picking at the substrate. For the first thirty minutes, however, the isolate individuals had trouble keeping up with the school's activities. This behavior seems to agree with Shaw's suggestion that, although isolated young fish (*Mendidia*) are "attracted" to the school, their ability to orient with respect to their neighbors is due to visual and locomotory development. Perhaps the isolate *Tilapia* were able to adapt more rapidly because of their prior contact with one another in their mother's mouth.

At the length of about 2 cm, animals occasionally break away from the main body and perform some activity alone, although within five minutes they rush back to the school.

The activities within the school are constantly changing. One animal rises to nip at the surface; if three or four animals follow, it is likely that the school will join them. However, another fish may head down to the bottom to nip at the substrate at the same time. The group follows one or the other, but rarely splits up. If three or four animals find themselves isolated, they swim hurriedly back to the school.

In a study using over twenty young schooling fish, Neutral was the coloration most frequently observed (96 per cent of the total observation time). Isolated animals, fish which had left the school for five minutes or more, and groups of up to ten animals, spent 5–27 per cent of the time in Neutral. The rest of the time they were Barred. If the experimenter was observing a group of animals in Neutral coloration they would come close to the side of the tank and stare back (pl. 4, *d*), perhaps in anticipation of food or simply out of curiosity. The Barred fish would invariably turn and swim away from the observer. Thus a school of ten or more animals appears less disturbed than smaller groups. Keenleyside (1955) states that "an isolated fish appears more restless and 'nervous' as shown by increased sensitivity to sudden disturbance."

Keenleyside has shown that when rudd (*Scardinius erythropthalmius*) and stickleback (*Gasterosteus aculeatus*) are presented with two schools they prefer the larger to the smaller group. This is true of *Tilapia* also, but Keenleyside's schools consisted of not more than twelve animals.

Noble and Curtis (1948) and Baerends and Baerends van Roon (1950) have found that schools of young cichlids increase in density and sink to the substrate if alarmed. This behavior was verified in my experiments.

An adult school remains in midwater, consistently heading in the same direction while swimming. The members engage in the same activities, such as feeding or nipping the substrate. The behavior of an adult school is far less homogeneous than that of a school of young because they are constantly being interrupted by the reproductively mature males and brooding females. Schools observed in aquaria were forced to remain above the nest areas and milled about, subjected to a barrage of courting, fighting, and chasing animals. For this reason the members of an aquarium school were usually Striped, whereas those in the pond were Neutral. In the pond, schools hardly ever passed over the nest areas. The members of these schools had access to substrate where no nests were situated. If a school ventured into the territorial areas, fighting and chasing ensued.

Under laboratory conditions, young up to 1 cm in length will be eaten by fish longer than 5 cm. This ratio of about 1 to 5 applies to other size groups until the small animals reach a length greater than 3 cm. At this size they are not eaten. Animals 3.5 cm long will be chased by a school of 6 cm fish, whereas 4 cm fish are accepted into the school.

In review, the young form tighter schools than adults. This is due in part to the interference in adult schools by courting and fighting males and brooding females. A minimum of ten young animals is needed to form a school in which all individuals are in Neutral and are not fearful of the experimenter. Young less than 3 cm long were eaten by aquarium-raised animals five times larger than themselves.

TERRITORIAL BEHAVIOR AND COLORATION IN THE MALE

Qualitative Analysis

Lowe (1959) and Ruwet (1963) state that in mouthbrooding cichlids other than *T. macrocephala* the male is solely responsible for the selection and construction of the nest. He also defends his territory against intruders. The behavior of *T. mossambica* supports this thesis. The first indications that a male will set up a territory are Dark 1 coloration, midwater courtship or threat behavior, and digging in a localized area. A male with an established territory is recognized by Black coloration, the lack of initiated aggressive display other than pendeling, and residence in the territory about 90 per cent of the day.

When two or more males are placed in a tank, both the process of taking a territory and the darkening of color take several days. Out of six previously isolated males introduced into a 70-gallon tank, one male established a territory. The data presented in table 1 show his color changes as he established a nest. Both territorial establishment and Black coloration were completed by the seventh day. Banded Black coloration was attained on the fourth day. In other observations of 3–4 cm young, which were assuming reproductive coloration for the first time, the Banded Black color developed in four to six days.

A simpler picture develops if a single male is placed in a tank with several females. Courtship and nest digging occur while the male remains in his territory. Within twenty-four hours he is Black.

Males which are establishing territories are extremely active, and their behavior is less predictable than that of resident males. The types of behavior most often seen are those oriented toward the part of the substrate chosen for the nest site. These are mouthdigging and interterritorial fighting.

TABLE 1
Percentage of Time in Different Color Patterns of Male 31 during Establishment of a Territory[a]

Day	Percentage of time in color state					Total time
	Neutral	Dark 1	Dark 2	Dark 3	Black	
4/12	34	62	14	..	..	30'
4/13	2	45	53	..	..	30'
4/14	..	5	73	22	..	30'
4/15	..	..	7	93	..	30'
4/16	..	..	..	86	14	30'
4/17	..	..	..	57	43	30'
4/18	..	..	..	..	100	30'
4/19	..	..	..	3	97	30'

[a] Data recorded for 8 consecutive days.

When only one male is placed in a tank he spends most of the time digging in his territory and courting the females. If other males are present they attempt to take over the nest. Fighting starts in one territory and gradually moves out into other parts of the tank. If two fighting animals move into another territory the resident male of that territory invariably joins in. As many as nine males were seen fighting in one corner of the tank. When fighting stops (usually owing to one fish biting or tailbeating the other) the nest-building fish heads back to his territory. After a few minutes the nest-builder is attacked by the same or another invader and fighting begins again until the latter is bitten or chased around the tank; the invader then either schools or transfers his attack to another nest-building male.

At this stage courting occurs rarely because there is little time for it. As more and more males are driven out of the area and do not return to fight, the winner begins courting frequently. During the territorial establishment phase, the males are Dark 3 most of the time. As fighting falls off and courting becomes frequent, the largest amount of time is spent in Black. After two or more territories have been established, the resident males remain in Black and do most of their courting in midwater.

If a strange fish of either sex enters a resident's territory and becomes Striped or Barred, Neutral, or Dark 1 it will be courted. If it becomes Dark 2, Dark 3, or Banded Black it will be threatened. If it flees it will be chased and consequently blanches to Neutral, Barred, or Striped. If it is a male which holds its ground, a fight ensues. Both males then become Dark 3. The winner chases the other away and occupies the territory.

Mouthdigging is one of the most common activities. By this means the territory is shifted constantly. All fish dug through the two inches of sand until the bottom of the tank was exposed. Then they shoveled away at the sides of the pits, spitting sand into adjacent nests. The more industrious animals enlarged their territories until other nests were gradually engulfed with sand.

Pendeling occurs between two resident males along the boundary between nests. A resident may frequent an unoccupied area between two adjacent nests and eventually drive the other male back to his nest. Finally, pendeling occurs on the border of the loser's nest. The winner digs another pit in the newly claimed area and presides over both nests.

A male rarely leaves his own nest to take over the territory of another male. One animal switched his territory to the left side of a 100-gallon aquarium when the fish were fed on the left side. When the feeding location was changed to the right side of the tank this same male regained his old territory on the right side. Both transfers were peaceful. The other two males moved into the abandoned territory both times without defending their own nests.

When a male approached a territory holder, the latter displayed only if the intruder was from 0 to 4 cm shorter than himself. Small males 3 to 4.5 cm long placed in a large aquarium containing males 6 to 12 cm long never succeeded in establishing territories. These small males were ignored by the resident males, although they became Dark 1 (a few reached Dark 2) and courted the large females in midwater. Two descended to the bottom and began to dig pits for the first half hour, but were chased out by the larger residents. Only one attempted to fight with a large male, but after one threat movement from the latter the youngster fled. After the first hour no further attempts were made to descend to the substrate except when they joined the larger males in their attempts to interfere with spawnings. Although smaller animals did not establish territories after ten days, three were Banded Black and courted in midwater.

If a resident's area is invaded by a swarm of animals less than 3 cm long, he chases them away by nudging as many as he can in quick succession. He often fails to drive them away and, if well fed, may allow the youngsters to poke about the nest and swarm around him. When the majority leave he nudges the rest out and continues his former activities. He may eat many of them if he is much larger than they are.

Quantitative Analysis

No conclusions can be reached about the possible motivational states of *Tilapia* without presenting a quantitative analysis correlating color patterns with behavioral activities. The results of this study represent 36 hours of observation on 54 animals.

Four groups of animals were used: Group I, 36 fish observed in the Steinhart Aquarium tank; Group II, 8 males in a 100-gallon tank; Groups III and IV, 6 males each in a 70-gallon tank.

The histograms summarize the time spent by males which are establishing territories and by resident males, in a particular color pattern while engaged in certain activities. The results of all four groups of fish are combined in figures 1 and 2. Fighting activities and courting components were lumped together with the exception of chasing, which occurs in both. Tailwagging, mouthdigging, and pendeling were put into separate categories because they presented special problems and had to be dealt with in more detail. Miscellaneous activities which were lumped together include swimming, resting, eating, nipping off any surface, entering and leaving the territory in a horizontal fashion (as opposed to tilting), yawning, rubbing on the substrate, fleeing, and rising.

Almost all courtship and fighting took place in the Dark 1 through Black series of color patterns. Males with established territories were either Dark 3 or Black and were not seen to lighten from these two colorations unless they lost their territories. When the color patterns of fish occupied in various activities were plotted, there was a predominant color pattern for each selected activity.

The time spent fighting in Dark 3 was much greater than the total amount of time spent in this color while performing all the other reproductive activities. The coloration for courting fish showed a slight peak at Black and a lower one at Dark 2 for territory-establishing males and a large peak at Black for resident males. The largest proportion of the total time in Black was spent courting. Mouthdigging in Black was similar to courting in its frequency histograms. Pendeling was performed only by Black resident males, although this activity is considered threat display. Tailwagging, the rarest of all the component activities listed in figure 1, occurred most often when the fish was Black. Chasing occurred most often when the territory-establishing males were Dark 2 and the resident males were Black.

Miscellaneous activities (fig. 2) were performed in all the color patterns. Animals with territories executed the majority of these activities in Black. Dark 3 was the most common pattern assumed by territory-establishing males when they engaged in these nonreproductive behaviors.

The Phi coefficients listed in tables 2 and 3 measure the correlation between the time spent in a color pattern and the time spent fighting, courting, and performing miscellaneous activities. Owing to insufficient data, chasing, tailwagging, mouthdigging, and pendeling were not included. Coefficients were calculated for each of the four groups of animals and there proved to be a high degree of consistency between them. Since resident males assumed only two color patterns, the correlation coefficients were higher for this group.

There is a good correlation between Dark 3 fighting males and Black courting fish for both territory-establishing males and resident males. A slightly less reliable correlation was found between Dark 3 fighting males and Dark 2 courting fish. There was random correlation between Dark 1 courting animals and Dark 3 fighting ones. Since the animal becomes Dark 1 at the start of either courting or fighting, no correlation was expected.

When fish become dark they perform most of their miscellaneous activities in these colors rather than fading to Neutral, Barred, or Striped. The latter colorations of fish performing miscellaneous activities thus showed a fairly low correlation to Dark 1 through Black fish engaging in reproductive activities.

The Rate of Color Change

From the data in the community studies it was found that *Tilapia* 6 cm long require only a few seconds to change from one color pattern to another. The time necessary for these changes was recorded on 140 occasions: 70 observations

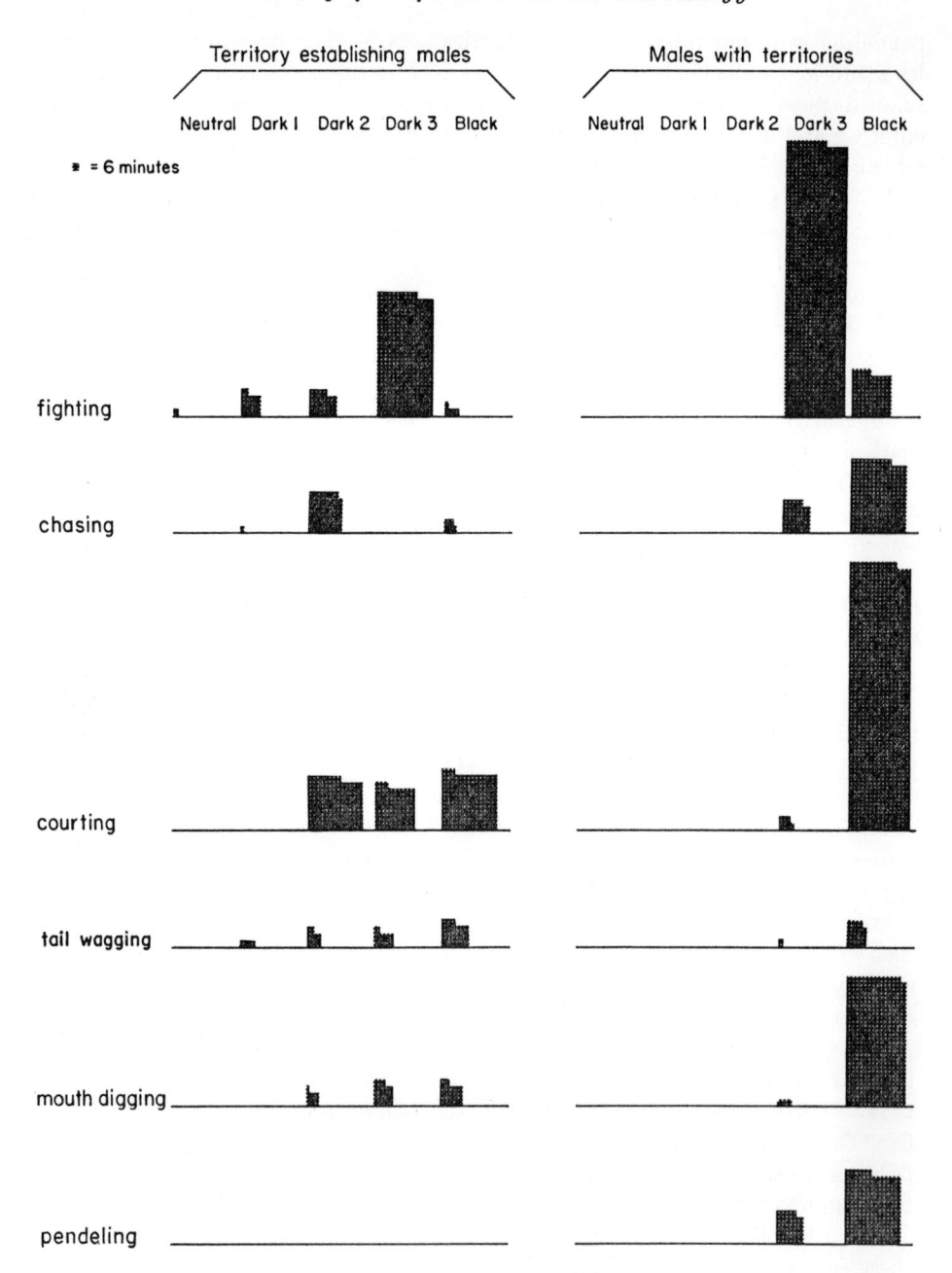

Fig. 1. Time spent in different color patterns during reproductive activities: 54 male *Tilapia* were observed for 36 hours, each for 20 minutes during territorial establishment and 20 minutes after having established a territory. The columns, based on a unit of 6 minutes, represent the amount of time spent in each activity in each color pattern, totaling 36 hours for all columns.

measured the time required to darken when a male was courting a female, and 70 measured the time required to fade when he was being chased. The number of seconds between color jumps that appear in table 4 are an average of the time required for the color markers of patterns to appear or disappear.

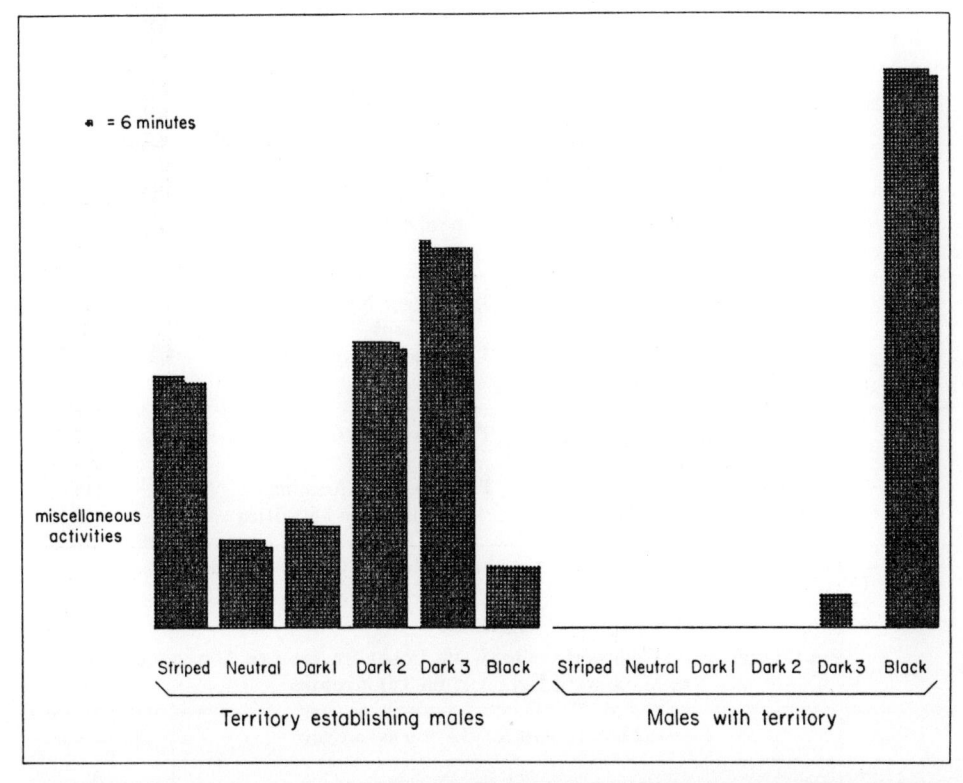

Fig. 2. Time spent in different color patterns during nonreproductive activities: 54 male *Tilapia* were observed for 36 hours, each for 20 minutes during territorial establishment and 20 minutes after having established a territory. The columns, based on a unit of 6 minutes, represent the amount of time spent in each color pattern, totaling 36 hours for all columns. Miscellaneous activities include swimming, resting, eating, nipping off any surface, leaving territory in a horizontal plane, fleeing, and rising.

AN ANALYSIS OF BEHAVIOR AND RESULTANT COLOR CHANGES

Since *Tilapia* change from one color pattern to another very rapidly, an attempt was made to correlate color changes with the previous action of territory-establishing males. The data recorded in figures 1 and 2 were used. The first ten minutes of observation on each of the 45 males is found in table 5, the second ten minutes in table 6. Table 5 shows the number of one-step color jumps of the 45 animals related to the behavior preceding them. Table 6 shows this same relationship, except that one or more color jumps from the initial behavior were used. Thus, if a fighting animal passed from Dark 1 to Dark 2 and then to Dark 3, a mark was placed in the Dark 3 column. When no color jump was recorded for a behavior, this was placed in the column describing the condition. Both tables have two dif-

TABLE 2
Phi Correlation[a] Coefficients between Color States and Activities in Male during Territorial Establishment

Group	Color and activity correlated with color and activity		Phi coefficient
1	Time spent in Dark 3	Time spent in Black	.651
2	and	and	.681
3	Time spent fighting	Time spent courting	.637
4			.662
1	Time spent in Dark 3	Time spent in Dark 2	.507
2	and	and	.557
3	Time spent fighting	Time spent courting	.420
4			.362
1	Time spent in Dark 3	Time spent in Dark 1	.01
2	and	and	.03
3	Time spent fighting	Time spent courting	.09
4			.12
1	Time spent in Striped, Barred,	Time spent in Dark 1 to Black,	.357
2	and Neutral, and	and	.298
3	Time spent performing	Time spent performing	.316
4	miscellaneous activities	miscellaneous activities	.362

[a] $\text{Phi} = (ad - bc)/[(a+b)(c+d)(b+d)(a+c)]^{1/2}$ where a, b, c, and d are frequencies of four variables.

TABLE 3
Phi Correlation[a] Coefficients between Color States and Activities in Male with Established Territories

Group	Color and activity correlated with color and activity		Phi coefficient
1	Time spent in Dark 3	Time spent in Black	.883
2	and	and	.875
3	Time spent fighting	Time spent courting	.820
4			.851

[a] See table 2.

TABLE 4
Average Number of Seconds to Complete Color Changes by 23 Males

	Color patterns							Total
	Hatched	Striped Barred	Neutral	Dark 1	Dark 2	Dark 3	Banded Black	
Darkening,[a] from left to right......	...	3.5	4.3	4.2	6.5	6.9	6.2	31.6″
Fading, from right to left.	7.3[b]	3.8	2.5	3.7	7.8	5.9	...	31.0″

[a] Darkening is measured in the courting male and fading in the fleeing male. Only animals that had previously been observed in Neutral were used in this experiment.
[b] To obtain this record the animal was prodded by the experimenter.

ferent sets of data, one for the behavior before darkening in color, the other for the behavior before lightening in color.

Tables 5 and 6 clarify much of the data in figures 1 and 2, owing to the following points:

1. Animals that were chased and left the territory became lighter in color. A Black male would blanch to Dark 3 during a fight and subsequently would fade even more if he was chased or left the territory. The Hatched pattern appeared only in animals that were chased. Chased animals or animals that were out of their territories very seldom failed to change color.

TABLE 5
ONE-STEP COLOR CHANGES OF 45 MALES OBSERVED FOR 4.5 HOURS
WHILE ESTABLISHING TERRITORIES

Behavior prior to darkening	Number of cases of darkening						Number with no change
	Striped Barred	Neutral	Dark 1	Dark 2	Dark 3	Black	
Being chased................	0	0	0	0	0	0	30
Fighting with male..........	0	25	105	206	411	0	341
Chasing male or female......	0	0	0	7	43	0	230
Courting male or female.....	0	16	73	79	292	437	645
Out of territory.............	0	0	0	0	0	0	0
Back to territory............	0	5	12	39	34	17	0
Resting, eating, schooling....	22	61	29	99	26	0	1,367

Behavior prior to fading	Number of cases of fading						Number with no change
	Hatched	Striped Barred	Neutral	Dark 1	Dark 2	Dark 3	
Being chased................	27	72	68	15	86	0	30
Fighting with male..........	0	0	0	0	9	267	341
Chasing male or female......	0	0	0	0	0	0	230
Courting male or female.....	0	0	0	0	79	0	645
Out of territory.............	0	29	38	37	62	0	0
Back to territory............	0	0	0	0	0	0	0
Resting, eating, schooling....	0	28	85	86	502	46	1,367

2. Fish darkened when they chased others, fought, courted, and returned to their territories. Fighting males never became Black. If initially Black, they faded to Dark 3. Much of the histogram registering Black fighting males in figure 1 represents the time spent fighting before changing to Dark 3. When courting, Dark 3 males often changed to Dark 2 rather than becoming Black or Banded Black. No Black male became Dark 2 while courting; all 63 color jumps were from Dark 3 to Dark 2. This decrease accounts for the fact that the Dark 2 histogram is larger than that of Dark 3.

3. Miscellaneous activities were combined in the tables as well as in figure 2. Most of these activities were pursued without color changes.

The results obtained from the studies on territory establishment as represented in tables 5 and 6 and figures 1 and 2 indicate that Dark 3 is associated mainly with fighting and Black or Banded Black with courting. Dark 2 is an alternative to Black as a courting coloration.

Cyclic Behavior of the Male

In any given period of time the number of males holding territories fluctuates. Forty males were studied to determine possible reasons for this fluctuation. Ten males were placed successively in a 100-gallon tank with six females and watched daily for three weeks. Three other groups were treated similarly. The color of each

TABLE 6
ONE- OR MORE STEP COLOR CHANGES OF 45 MALES OBSERVED FOR 4.5 HOURS
WHILE ESTABLISHING TERRITORIES

Behavior prior to darkening	Number of cases of darkening						Number with no change
	Striped Barred	Neutral	Dark 1	Dark 2	Dark 3	Black	
Being chased...............	0	0	0	0	0	0	97
Fighting with male..........	0	8	42	11	327	0	190
Chasing male or female......	0	0	0	3	15	0	254
Courting male or female.....	0	0	22	62	102	187	446
Out of territory.............	0	0	0	0	0	0	4
Back to territory............	0	0	0	9	8	17	11
Resting, eating, schooling....	0	50	35	38	6	0	687

Behavior prior to fading	Number of cases of fading						Number with no change
	Hatched	Striped Barred	Neutral	Dark 1	Dark 2	Dark 3	
Being chased...............	21	54	18	10	43	0	97
Fighting with male..........	0	0	0	0	0	155	190
Chasing male or female......	0	0	0	0	0	0	254
Courting male or female.....	0	0	0	0	63	0	446
Out of territory.............	0	17	0	0	12	0	4
Back to territory............	0	0	0	0	0	0	11
Resting, eating, schooling....	0	22	56	29	101	18	687

male was recorded once an hour during the day. The number of Black males was averaged to the nearest 1.0 per day. Eight of the largest males, two from each group, were tagged. (See table 7.)

In all four experiments the number of Black males and territories increased slowly during the first week. In the second week the number of Black males and territories fluctuated, and by the third week they remained constant. In the last week, with the exception of one day, three males held territories and no others were constructed. Each tagged animal that established a territory remained in it for only part of the three weeks. This period of time appeared to be different for each animal. The length of time spent schooling was fairly constant for the tagged animals, varying from six to eight days.

To my knowledge, only a few experiments have been conducted on cyclic behavior among fish under controlled lighting and temperatures. Various authors do not concur as to the causal mechanism of this behavior. For example, Brawn (1961) noted alterations in the dominance of *Gadus callarias*. She felt that this depended on the ability of cod to withstand threat. Maruyama (1948) studied the effects of temperature and day length on the activities of *Tilapia* and found definite changes in these two variables. With such a limited amount of data with respect to my own

TABLE 7

CHANGES IN NUMBER OF BLACK MALES AND NUMBER OF TERRITORIES IN FOUR EXPERIMENTAL TANKS[a]

Day	Average no. of black males	Average no. of territories	Group 1 A	Group 1 B	Group 2 A	Group 2 B	Group 3 A	Group 3 B	Group 4 A	Group 4 B
1	2	1	..	X	..	..	..	..	X	..
2	1	1	..	X	X	..	..	..	X	..
3	1	1	..	X	X	..	..	X	X	..
4	1	1	..	X	X	..	..	X	X	X
5	2	2	X	X	..	..	..	X	X	X
6	3	2	X	..	..	..	..	X	..	X
7	3	3	X	..	..	..	..	X	..	X
8	4	3	X	..	X	..	..	X	..	X
9	7	4	X	..	..	..	..	X	..	X
10	3	3	X	..	..	..	..	X	..	..
11	4	4	X	..	..	..	..	X	..	..
12	5	3	X	..	X	..	..	..	X	..
13	8	4	X	X	X	..	..	..	X	..
14	3	3	X	X	X	..	..	..	X	..
15	3	3	..	X	..	..	..	..	X	..
16	3	4	..	X	..	..	..	..	X	X
17	3	3	..	X	..	..	..	..	..	X
18	3	3	..	X	..	..	..	..	..	X
19	3	3	..	..	..	..	..	..	..	X
20	3	3	X	..	X	..	..	..	..	X
21	3	3	X	..	X	..	..	..	..	X

[a] Each tank contained 10 males and 6 females. Continuous records of 8 marked males when they held territories.

animals I am not prepared to discuss whether internal or external factors play the major role in the activity of male *Tilapia*. The fact remains, however, that territorial behavior is probably cyclic among *Tilapia*.

SEX DISCRIMINATION AS RELATED TO MALE COLORATION AND BEHAVIOR

Twenty male fish were placed in a 100-gallon experimental aquarium and the same number in a 70-gallon tank to serve as controls. Four experiments, each lasting four days, were followed by a four-day rest interval. In each experiment the number of Black males was recorded every two hours and averaged to the nearest 1.0. First, both groups were observed alone. Second, one liter of water was introduced into the experimental tank from the female stock tank. The same amount of water from a male stock tank was poured into the control tank. Third, a 15-

gallon aquarium containing ten females was placed next to the experimental aquarium and a similar tank containing ten males was placed next to the control tank. Fourth, ten females were placed in the experimental tank and ten males in the control aquarium.

Only the last experiment produced a significant increase in the number of Black fish. In the first three experiments, from 0 to 3 males were Black in the experimental and control tanks. When the females were introduced, 9, 8, 11, and 17 males were Black on days 1–4, respectively, whereas 0–3 males were Black in the control tank.

In two further experiments, four females were placed on one side of a partitioned aquarium and four males on the other. The partition was opaque in one experiment and transparent in the other. There was free passage of water throughout the tank. In each group of males one became Black, but the rest remained Neutral for seven days. This seems to indicate that visual and chemical stimuli received from the female separated from the male are not sufficient to induce color change in the male.

In the experiments above the behavior of Black males was not recorded. The introduced males, however, seemed to be courted as often as the females. To determine whether this observation was correct, the male *Tilapia* were left in a tank for one week. The behavior of these resident males was recorded on the eighth day, immediately after the introduction of five males. These five males were taken out a few hours later and five females were inserted on the fifteenth day. There was no significant difference in the response of resident males to new males or females. In comparison with figures 1 and 2, table 8 shows a large increase in the amount of time engaged in reproductive activity.

Another set of experiments was conducted in order to determine whether the schooling behavior and coloration of introduced males were relevant factors. Ten males were placed in a tank for seven days (table 9). In every observation period the resident males tended to tilt to the new animals rather than the test animals left in the tank from the previous week (table 9). In the two experiments where new males and females or familiar males and females were observed, females were courted more often than males. All the test animals were Striped, Barred, or Neutral, and schooled continuously.

The territory-holding male courts the unfamiliar fish more often than the familiar, regardless of sex, and the female is tilted to more often if both sexes have been exposed to the resident male for the same length of time. The resident male behaves similarly with males and females if both sexes have been in the same tank with him for one week, although the frequency of courtship is in favor of the female. The time at which the sex of new animals is recognized was not determined.

COURTSHIP BEHAVIOR AND COLOR CHANGES

To determine if the male needs prior experience with a female to show all the components of courting behavior, 20 isolate male fish were used. The females used were taken from a stock tank that contained members of their own sex. Each experimental male was placed in one half of a partitioned 15-gallon tank and a gravid female was placed in the other half. After three days the opaque partition was removed. These experiments were compared with similar ones using 30 males that had been raised in groups. The following information is applicable to both isolated and community raised males unless the term "isolate male" is used specifically.

TABLE 8

BEHAVIOR OF TEN RESIDENT MALES AFTER INTRODUCTION OF
NEW MALES AND NEW FEMALES INTO THEIR TANK

	Percentage of total time engaged in activity				
	Courting	Mouthdigging	Fighting	Chasing	Resting
5 males[a] introduced	21	20	13	13.5	33
5 females introduced	28	20.5	20	14.5	17

[a] Each male was observed for 30 minutes; total observation time was 5 hours.

TABLE 9

TILTINGS BY TEN TERRITORY-HOLDING MALES TO FAMILIAR AND
NEW MALES AND FEMALES[a]

Test animals		Average no. of tiltings in 15 minutes		No. of days after introduction of original males
Left in tank	Introduced	To males	To females	
	5 new males 5 new females (both sexes tagged)	111 ± 17	123 ± 5	8
5 tagged females	5 males	197 ± 4	134 ± 11	16
5 males	5 females	4 ± 3	321 ± 3	24
5 males, 5 females	......	11 ± 2	108 ± 6	32

[a] Each resident male was observed for 15 minutes in each of the four experiments.

INITIAL BEHAVIOR AND COLOR CHANGES OF ISOLATE MALES

Isolate males were observed for three days before experimentation. When placed in the tank, the males were Striped, Barred, or Hatched and either dashed about or rested motionless with all fins raised. Within two hours to two days, 13 males were Neutral; seven were Dark 1 and began to dig pits.

All 20 males courted when they saw the female for the first time. The first tilt took place 57 ± 28.4 seconds from the time the partition was lifted. Darkening of color occurred about 15 seconds earlier than the first tilt. Initial behavior and color changes of community-raised males were the same.

QUALITATIVE ANALYSIS OF COURTSHIP

Courtship of *Tilapia* represents a series of half-completed acts depending for completion on the behavior of the female. If she responds correctly to the male, for instance by halting when he tilts, the male's behavior changes from tilting to leading. After the male tilts, the female may swim away; the male may then perform a prolonged series of tilts or he may tailwag and carry out another series of tilts.

If the female halts but does not follow the male down to the nest, the male begins twitch-leading the female or he may tailwag. Failing in a series of these actions, he may lead down without her and begin tilting again. Should she follow him down to the nest, the male then rolls (pl. 4, *b*). Rolling is followed by tilting and

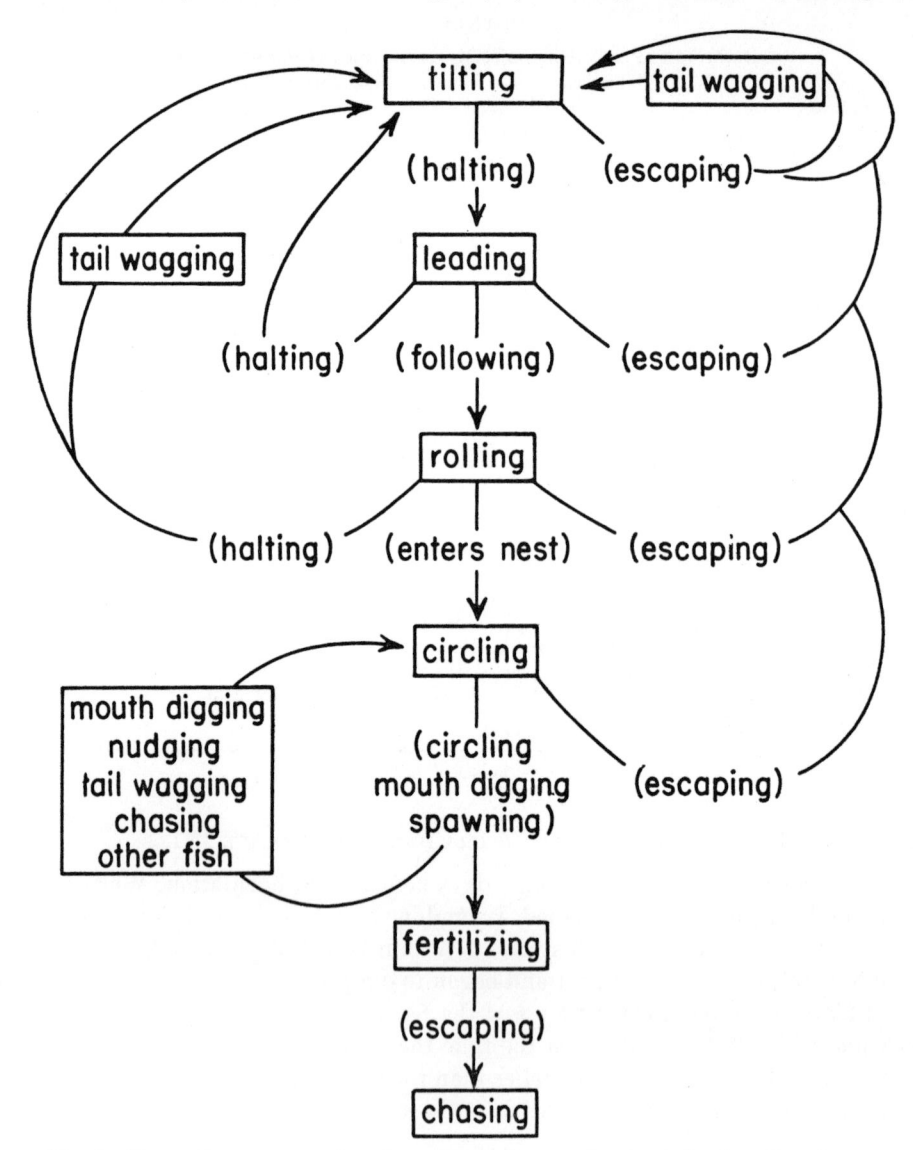

Fig. 3. Observed sequences of male and female reproductive behavior (female components in parentheses).

leading when the female does not enter the nest. If the female enters the nest, she and the male circle together. As she circles, he may break off and perform a number of courtship activities: tailwagging, twitching, nipping her genital pore, and mouthdigging. These activities continue until spawning occurs (fig. 3).

INFLUENCE OF THE FEMALE ON MALE COURTSHIP

Figures 4–6 are difficult to understand without reference to the female behavior associated with courtship. In the early stages of courtship and until they have dug in a nest, females normally remain in midwater until they are led down to the substrate. Tilting and twitch-leading or leading occur when the female is in midwater, whereas rolling occurs when she is very near the territory. With fish 7 to 11 cm long, the response of the male is correlated with the distance of the female from the nest (table 10).

TABLE 10
VERTICAL DISTANCE OF FEMALE FROM SUBSTRATE AND
CONSEQUENT ACTION OF MALE

Female	Action of male	
No. of cm above substrate	Frequency of rolling	Frequency of tilting, twitching, and leading
0–3	189	0
4–8	121	11
9–14	7	135
15 and above	0	310

During courtship the male tries to orient himself laterally with respect to the female, thereby exposing his striking courtship coloration and white genital papilla. Wickler (1962) suggests that the white papilla of the male *Tilapia* is a stimulus (egg dummy) to induce the female to approach. He is unable to do this when he is on or near the substrate of the territory; therefore he rolls over in the nest, exposing his lateral profile. When the female is within an arc 15 cm or more above the substrate or is on sand outside a 20 cm circle from the nest, the male tilts and leads her toward the nest. When the female is within these boundaries, the male rolls. I have reservations about whether the exact measurement of the arc determines male behavior, for, of the 20 males studied, only 7 dug nests prior to the encounter; of the remaining 13 males, 10 chose a spot and used it as if it were a real nest. The latter arcs were less accurately measured.

Figure 4 was constructed from observed sequences of 20 isolate males from the time they first courted the female until she dug in the nest. The majority of sequence breaks came after tilting. Out of 743 leading-tilting sequences, 176 were initiated by fleeing females. A male would rush after the female, lead, and begin tilting again. The other 567 lead-tilt sequences occurred in a similar fashion but with other acts interposed. A break in this chain was most commonly instigated when a female fled from a tilting male. When the male caught up with the female he would lead down to the nest as if the sequence break had not occurred, turn back, and tilt. This usually started the whole backward sequence over again. Occasionally a male would lead down toward the female and tilt as he passed her. Tailwagging occurred only after tilting. After tailwagging, the male either swam off or began tilting again. In the eight sequences where rolling followed tilting, the female was directly above the nest. Although there is a suggestion of progression from tilting to leading to rolling (see fig. 4), this is obscured by the tendency of the female to flee rather than halt.

Figure 5 shows the courtship sequences of the male when the female is near the nest or is on the substrate. If the female moves away from the nest area, the male stops rolling and tilts instead. Tailwagging nearly always seems to drive her off.

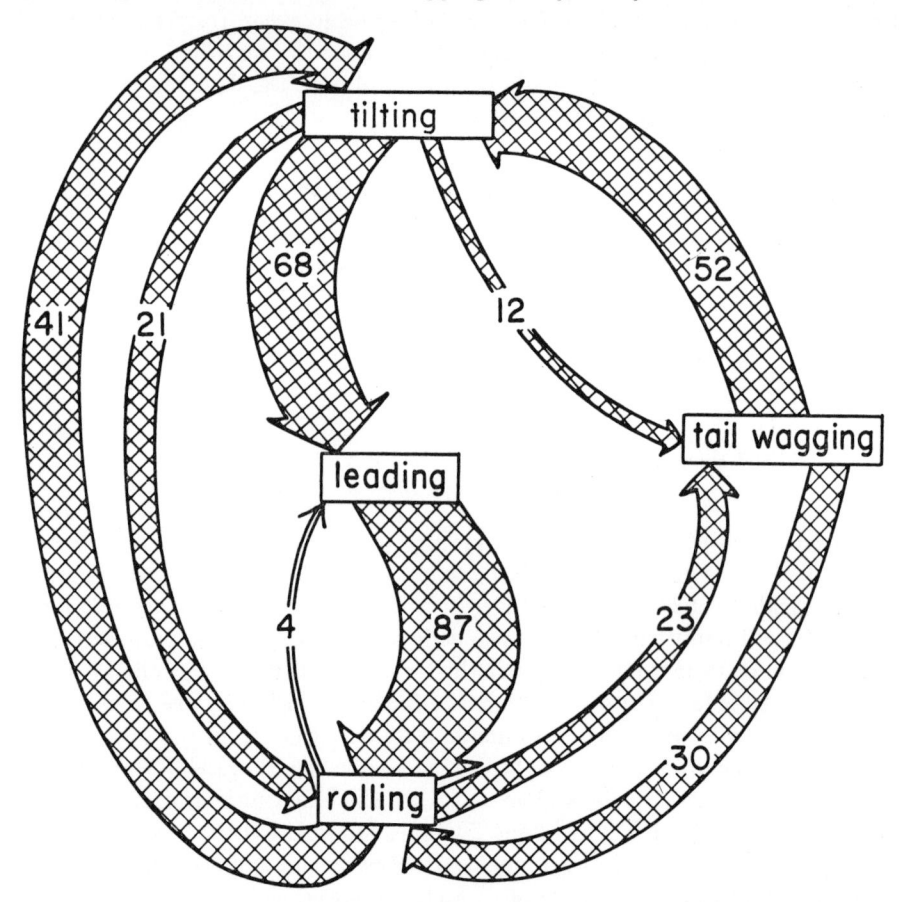

Fig. 4. Courtship activity sequences of male when female is in midwater, 15 or more cm from nest. Taken from data on 20 males observed for 5 hours. Only sequences of two or more activities were used. Where rolling occurred, the sequence was terminated.

Sequences of rolling-tailwagging-tilting-leading and of tailwagging-rolling-tilting-leading are common. In the latter sequence, tilting and then rolling occur when the female is close to the nest.

In summary, if the female is courted in midwater she tends to flee and thus the tilt-lead sequence of the male occurs numerous times until she comes down to the substrate. Approximately ten times as many of the tilt-lead sequences take place in midwater as on or near the substrate (cf. figs. 4 and 5). When the female is on the substrate she needs little coaxing to respond to tilting, leading, or rolling.

When the female engages in digging, she always completes spawning. Females enter and leave the nest area three or four times before they settle down to dig.

The males go into midwater to bring them down during this interval. These later tilt-lead-roll sequences grow shorter in duration until the female re-enters the nest without active inducement by the male. If the male is in attendance she often precedes him to the nest.

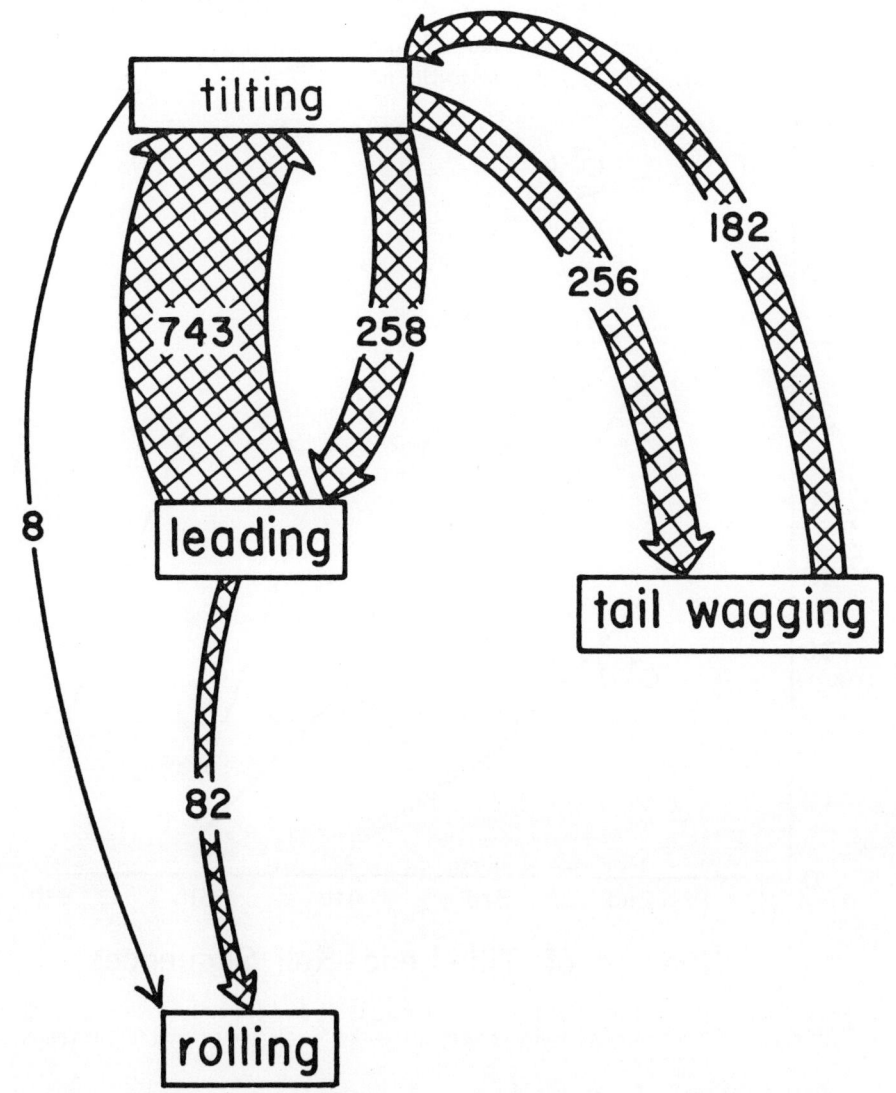

Fig. 5. Courtship activity sequences of male when female is on substrate or within a 15-cm arc above substrate. Taken from data on 20 males observed for 5 hours. Only sequences of two or more different activities were used.

TAILWAGGING

In compiling the data from all the observations made on male-female encounters, I have defined "more active" courtship as sequences consisting of more than eight changes in a male's activity per minute. During periods of more active courtship one or more tailwags are given. Among the 20 male fish, the average number of tailwags was 0.5 for the first minute, and 1.9 and 1.5 for the second and third minutes after his initial introduction to the female. After this and until the time the female entered the nest the number of tailwags fell to 0.05 per minute. During a sample of three minutes during which the female was close to the nest and within it, the number of tailwags increased to 0.45, 1.65, and 1.05. The number then dropped again. Tailwagging occurs more frequently when a male meets a female for the first time and when she is close to the nest.

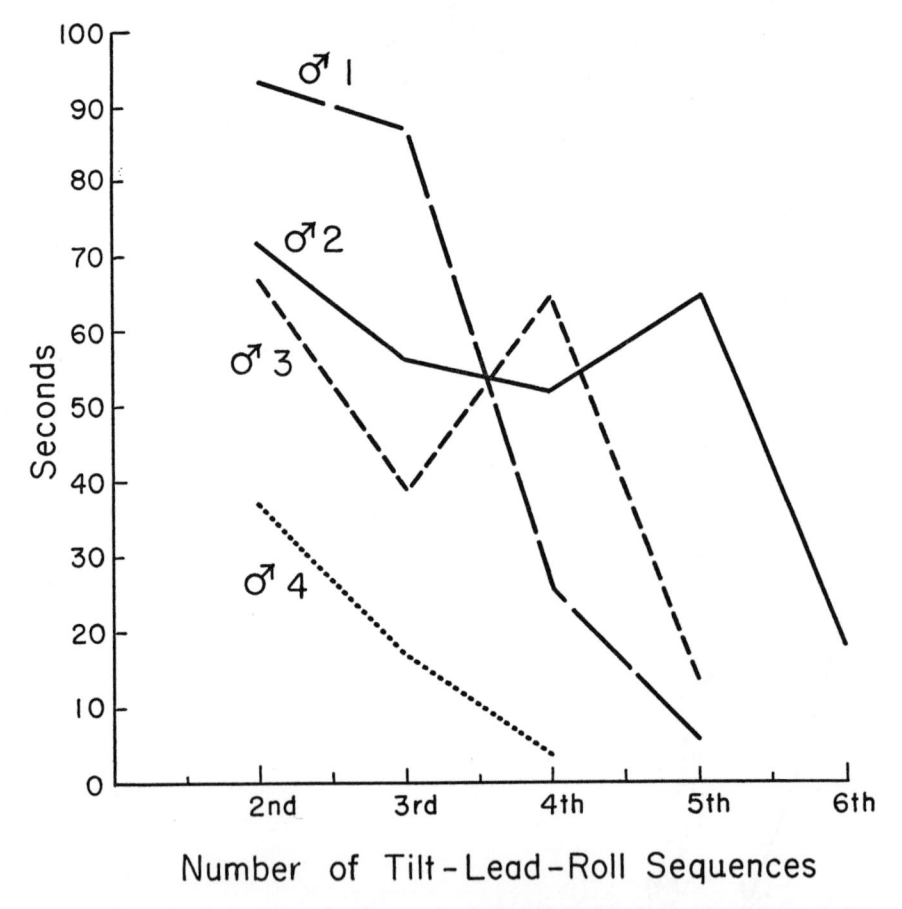

Fig. 6. Length of time spent by four males luring their females back to the nest. This series of consecutive attempts by males occurred after the female had entered the nest once.

If a female moves at all after the male tailwags she moves in the opposite direction. This is evident from figures 4 and 5, in which the male sequence is tailwagging-tilting or tailwagging-rolling-tilting. The female was observed more closely when the male tailwagged. Using the data recorded for 30 females it was found that they were in midwater and motionless for more than ten seconds before 965 out of 1,126 bouts of tailwagging began. Similarly, when the females were about to enter the nest but hesitated for ten or more seconds, 245 out of 282 bouts of tailwagging occurred. When the female moved away, tilting initiated a new sequence. When the female hesitated near the nest, after the male tailwagged he either tilted or rolled, depending on her position.

The male's more active courtship appears to result in immobility of the female, especially when she must either follow him or flee. If she does not move when he attempts to lure her toward the nest, he tailwags in order to get her to move so that he may resume courtship.

INFLUENCE OF THE MALE ON COURTSHIP

Until now the effect of the female's behavior on male courtship has been emphasized. A male may alter the speed of spawning also: he may rest or swim around the tank, thus lengthening the period of courtship (table 11).

TABLE 11
PERCENTAGE OF TIME OCCUPIED BY 20 MALES IN COURTING FEMALES

Percentage of time courting	Number of males
1–10	1
11–20	3
21–30	2
31–40	3
41–50	3
51–60	3
61–70	2
71–80	1
81–90	1
91–100	1

Owing to the extreme variability of time spent courting and the small number of animals used, isolate and community males have not been compared. Instead, the average number of courtship activities of isolate and of community males has been compared. The number of changes in courtship activity used in this comparison were very large. There are 3.7 ± 0.93 such changes per minute for isolate males and 2.6 ± 0.41 changes per minute for community-raised males. The isolate group, when retested after several weeks of living in a community, performed 2.3 ± 1.0 changes in courtship activity per minute. Although there is a fair amount of individual variation in both groups, isolate males were more active on their first encounter with a female than were males with prior experiences.

TEMPORAL RELATIONSHIPS BETWEEN COURTING AND MALE COLOR CHANGES

In both the community and isolate studies on mature male *Tilapia*, courtship activities and the darkening of body color invariably occur together. In a Neutral male, darkening continues until the time of spawning, when the male is either Dark 2 or Banded Black. At the end of the experiments with isolate males the color of six of the largest males was Banded Black and the others were Dark 2. No animal was Dark 3 and two had faded from Dark 3 to Dark 2. Males that do not become Black after about thirty minutes fade to Dark 2 and court in this coloration. In each of the eighteen isolate encounters, tilting preceded and was closely related in time to the adoption of the Dark 1 color pattern. From Dark 1 on, the animals approached Banded Black as they went through the various stages of courtship; however, no particular color pattern was necessarily associated with any action.

CONCLUSION

Isolate male *Tilapia* court females in the same fashion as community-raised males. This observation is not unique to this species. Wiepkema (1961) also observed all the reproductive movements in 80 isolated *Rhodeus*. Isolate and community-raised *Tilapia* perform the majority of courtship activities in a manner dictated by female behavior. Isolate males execute sequences of activities more rapidly than community-raised males. All color patterns correlate with the courtship activities as portrayed in figure 1.

PRESPAWNING AND SPAWNING BEHAVIOR

QUALITATIVE ANALYSIS

A ripe female assumes a dark silvery coloration (pl. 4, *b*). Until she is ready to spawn she circles with the male in the nest and mouthdigs. Both male and female leave the nest for various reasons. During prespawning and spawning, other males frequently attempt to enter the territory. If a fight develops the female flees. After an interval she comes down from the school to one of the nests. As time progresses her visits are confined to one territory, where she joins the owner in defending it. When the female joins a male in the nest he circles with her, leaving momentarily to chase away neighboring fish. He returns in the tilting posture and resumes prespawning activities. Even in the absence of other fish in the area, the male occasionally leaves the nest, rushes about the tank, and returns.

The female may periodically vibrate without liberating eggs. This is called an "unsuccessful spawning attempt." When the female lays her first batch of eggs, the male places himself behind her and may eat many of the eggs. After the female has laid her first eggs and is collecting them, the male may repeatedly nudge her. If he bumps her too often or too hard she leaves the nest and goes to another territory to spawn. Lowe (1959) and Ruwet (1963) observed successive polygamy and polyandry among the *Tilapia* species.

QUANTITATIVE ANALYSIS OF FEMALE BEHAVIOR

A number of ripe females were studied in a search for the basis of their choice of spawning partners. Continuous observations were made on one prespawning female from the time she first dug in a territory (table 12). The observations were made in a 100-gallon tank containing ten males and ten females. This is typical of spawnings in the presence of other males, except that it was the shortest one recorded.

The largest amount of time was spent in the territory in which the female mouthdug, helped defend, and eventually spawned. All the territorial males were visited by the female when she changed position. The female fled to midwater only when a fight began.

Table 13 presents data on the prespawning and spawning activities of six

TABLE 12

PRESPAWNING AND SPAWNING ACTIVITIES OF ONE FEMALE
FROM TIME SHE FIRST MOUTHDUG IN A NEST

Female I	Female in territory					Female in midwater	Total
	B2	B17	2S	7L	3M*		
Total no. of minutes...........	9	89	20	1	1	54	168
Total frequency...............	5	7	7	1	1	10	31
Return from midwater.......	4	2	3	1	..	..	10
Change to specific territory..	1	5	4	1	..	..	11
First mouthdig................	..	×	..	..	..	..	..
First help in defense..........	..	×	..	..	..	..	..
First vibration................	..	×	..	..	..	..	..
First spawning................	..	×	..	..	..	..	..
Number of minutes spawning...	7	30	7	19	..	..	63
Frequency of spawning........	1	4	1	2	..	..	8

These activities are not in sequence but are categorized in groups showing position of female in aquarium. Resident males were tagged and the territory they occupied bore their label. An asterisk indicates a temporary territory.

TABLE 13

LOCATION OF SIX FEMALES IN TERRITORIES OF MALES DURING
PRESPAWNING AND SPAWNING

Aquarium	Female	No. of males with territories[a]	First occurrence of female prespawning activities with males A-H			Spawning sequences with males A-H	
			Vibr.	Defense	Cont. mouthdig	Sequence[b]	No. with eggs laid
100 gal............	5	4 A-D	B	B	B	B-A-C-B-B-D*	5
100 gal............	6	4 A-D	C	A	A	A-A-B-A-C*-A*	4
100 gal............	8	3 A-C	C	B	B	B-C-C-B-B*	4
100 gal............	21	3 A-C	A	A	A	A-C-A*-B-A*	3
70 gal............	3	3 A-C	A	..	C	C-C-C-C*-B*-A*-C-C*	4
Steinhart..........	14	8 A-H	E	F	F	F-G-F-F-E*-E*	4

[a] Different males were used in each case, although they were labeled identically.
[b] Asterisk indicates unsuccessful spawning attempt.

females, selected because each chose a different resident male for the first spawning. Since the first mouthdigging of four of the females was not observed, continued mouthdigging in one territory was adopted for the start of continuous observation. The six females spawned sixteen times with males in whose territories they continuously mouthdug and which they helped defend. Eight additional spawnings occurred with males other than the chosen partner. There were as many unsuccessful spawning attempts with their partners as with other males.

The first spawning vibration occurred only twice in the territory where the female continuously mouthdug, assisted in defense, and spawned for the first time. Four females spawned with all resident males in the tank; the other two spawned with males in the territories adjoining the one in which they laid their first eggs.

THE NATURE OF FEMALE BEHAVIOR

Baerends (1950) wrote of the female *Tilapia*: "... only a few hours before spawning she is attracted by one special male or possibly only by its territory which she even helps defend." It is evident from my research that the female makes a choice of spawning location. Is she attracted by the male, his territory, some individual characteristic or color he has adopted at the moment, or combinations of these factors?

Resident males cannot usually be induced to change territories. On one occasion a female had been mouthdigging for twelve minutes in the nest of resident male B5. When C3 (another resident male) entered, the female and B5 were chased away. The female quickly returned to B5's old territory, helped C3 defend it, and spawned with him. She ignored B5, although he was near the nest. Mouthdigging for twelve minutes seems to have created a bond between the female and the location. Females dig in several territories. As a female spends more and more time digging in one area, the other nests are dug infrequently; if chased into midwater from her chosen nest, she often performs digging movements at the sides of the aquarium.

Females tend to dig with the male that has the largest territory. The Black coloration of male B17 lacked white head markings. This male continually had a large territory, usually more than half of the 100-gallon aquarium. During the nine hours in which he was observed, four females dug persistently in his territory and he spawned eight times with them. Two of these females initially prespawned with three other resident males, which spawned ten times in the nine-hour interval. The larger the territory, the less interference is encountered from neighboring males, and the female was able to remain in the territory during a fight. Furthermore, if Black coloration of the male were a prerequisite for attracting females, B17 should have been visited by fewer females. The Black color pattern is not necessary to evoke a response from the females, for many spawned with Dark 2 males and even Dark 3 males (e.g., males in Dark 3 which had come down from midwater to enter into the spawning activities).

A single experiment demonstrates that a partner is needed before the female will spawn.[2] After a female had spawned her first batch of eggs, an opaque partition was brought down between her and the male. Although the female continued to circle the nest and hunt for more eggs, she did not spawn for two hours. When the partition was raised again she spawned within 25 seconds after seeing the male.

In conclusion, three factors seem to be needed by a ripe female in order that she may spawn. First, a nest occupied by a dark courting male will attract her down to the pit. The male's white genital papilla might be of great significance in leading the female (Wickler, 1962); however, my research was conducted prior to Wick-

---

[2] This technique was tried again and again, but the fish were badly frightened thus accounting for the single success.

ler's publication and this was not checked thoroughly. Second, as she continuously digs in the nest, a bond is established between her and the territory; so, if displaced, she returns to the same nest. Third, a dark male must be near or in the pit when she spawns. Larger territories, with less interference from other males, favor successful spawning.

BROODING FEMALE BEHAVIOR AND COLOR CHANGES

Since there was no place to hide in the community tanks, brooding females tended to swim up and join the school for the first week. If isolated or if in a school, they assumed the Neutral or Striped coloration (pl. 5, *d*). At the end of the week the similarities between isolated brooders and those in the community tanks ended. During the second week, the brooders in the community tanks became Hatched with dark eyes and a dark mouth region. They tended to remain in one corner and defend it by rushing at the intruder, by lateral displays, and by tailbeats (pl. 5, *c*). Animals in isolation also took up residence in one area, but remained Striped with dark eyes and mouth. The isolated female on spitting out her brood remained in Striped coloration with a dark mouth and eyes (pl. 5, *b*). In the community tank, when the young were spat out for the first time they were eaten by the other fish. The mother would then station herself by the air hose and defend the bubbles with extreme pugnacity. Between attacks on other fish, she would try to scoop up the bubbles in her mouth or would make "calling movements" to them. Brine shrimp were treated as young and would be brooded for some minutes, spat out alive, caught up, and brooded again. This behavior continued for about three days, in Hatched coloration. Finally, either hunger or some other factor or both brought it to an end and the female joined the school again.

Response of young cichlids to the color of their parents depends on species coloration. Baerends and Baerends van Roon (1950) found that very young *Tilapia* were attracted to the dark eye and mouth region of their mother and would attempt to penetrate the dark areas of black and white models. Noble and Curtis (1939) found that young *Hemichromis* would follow red models, but would scatter if a black model was introduced. Young *Cichlasoma biocellatum* would follow black models. It was suggested by Kühme (1962) that the preference of young *Cichlasoma* for black objects was possibly "shelter-seeking" and homologous to the dark-region seeking of *Tilapia*. Reversal of an attractive color pattern should therefore produce a different reaction from the young. With *Tilapia*, after six to ten days, three-fourths of the isolated mothers became Neutral again and one-fourth changed to a different color pattern. The latter had a light lower jaw, opercle, and belly; the rest of the body was dark. When these changes took place, the young either paid no attention to their mother or swam to the darker parts of her body. Concurrently, the female "called her young" infrequently and spent most of her time swimming about. Both the color and the activity of the female at this time indicated that she was ready to leave her offspring and join the adult school again.

In conclusion, the color patterns of brooding *Tilapia* are correlated with the readiness with which they become alarmed. About seven days after spawning, the young begin to wriggle in the mother's mouth. At this time she becomes Hatched in the community tank, or Striped when isolated. Possibly there is some connection between movement of the young and the Hatched color. When the young are eaten by other fish, the mother, after trying to defend them, finally collects the remaining few, becomes Hatched, and attempts to defend and snap up bubbles coming from the air hose. I feel that the female may become Hatched when guarding air bubbles because she is not able to gather them up and brood them, and hence exhibits this coloration. When a female reverses her brooding color pattern or becomes Neutral again, her behavior and the behavior of her young indicate that both the female and her young are ready to school separately.

FIGHTING BEHAVIOR AND COLOR CHANGES

The data on fighting behavior came from two sources: 44 males isolated for one week after emerging were used to determine whether mature males respond to their first contact with another male in the same manner as they do in groups of fish living together from birth.

One male was placed in each half of a partitioned 70-gallon tank. On the fifth day the opaque partition was lifted. If no fighting occurred after one hour, constant observation was discontinued. If fighting took place, continuous recording was terminated when one animal fled from the other and became Neutral, Barred, or Striped. Spot checks were taken hourly throughout the rest of the day. The animals were removed from the tank on the following day and were replaced by two new fish.

Behavior and Color Changes of Isolate Males prior to and during the Initiation of a Fight

When placed in the experimental tank, the 44 males took from three hours to two days to adapt to their surroundings. During this time the animals were Striped, Barred, or Hatched, and either dashed about or rested with all fins raised. After some time they began to swim more slowly and rested with their median fins closed. Most of the males became Neutral, but a few became Dark 1 and began to dig pits. Similar behavior was observed with community-dwelling males.

When the partition was lifted the two isolate males usually sank to the floor of the tank. Of the 44 animals, 16 became Hatched as a result of the disturbance. The others remained in Dark 1 and Neutral or assumed the Striped pattern. In all experiments the first animal to notice the other immediately joined him on the substrate. The average time taken to do so was 27 ± 12.3 seconds.

In fifteen encounters the initial reaction of one male was to court the male which crossed into his territory. In four encounters, one of the two courted throughout the hour of observation. In eight of the remaining eleven, the courting male followed his opponent into the latter's part of the tank and was threatened; in the other three, the courting male was threatened in his own territory. Fighting took place within three minutes after one animal joined the other.

In the other seven encounters, threat display and fighting commenced within 60 seconds after one fish joined the other. The intruder was attacked and no courting was observed until the fighting ceased. The length of time between the start of an encounter and fighting averaged about 32 seconds. An average of 63.6

seconds elapsed from the time one fish joined the other, was courted, and then threatened. The 22 males which courted or initially threatened assumed the Dark 1 color phase in about 75 seconds. The rest remained in Hatched, Striped, or Neutral until they began to threaten their opponents. In these animals, the color jump to Dark 1 took about 90 seconds.

FIGHTING BEHAVIOR: A QUALITATIVE ANALYSIS

Male *Tilapia*, whether isolate or raised in groups, do not fight continuously but confine their aggressive activities to a series of bouts. During the intervals between bouts of fighting the animals rest, court, or swim about.

A bout normally begins with a lateral display and tailbeating (pls. 3, *d*, 4, *a*). Preliminary tailbeats are performed two or more inches from the opponent so that no physical contact is made. Frontal display occasionally precedes a fighting bout, but mouthfighting was rarely observed in these early stages. When the other animal retaliates with a lateral display and tailbeats, fighting starts in earnest. The two fish begin to carousel upward, each attempting to place himself slightly above and lateral to the other. Upon reaching this position, one of the fish may hurl himself downward and laterally, slapping the side of his opponent with his caudal fin. Upon reaching the substrate the two may carousel again. The ascent is usually accompanied by butting and biting at the belly, flanks, lower fins, and head region of the uppermost fish. Biting is rare in fights, but very common when a male chases or attacks a motionless fish. Although mouthfighting is not common, this method of attack is dramatic and powerful. The fish move some distance away, turning rapidly. From a frontal-display position the two rush at each other with open mouths. When they near one another, one of three things may happen. The two may display frontally, an inch away from each other, without biting. More commonly, they may attempt to mouthfight, fail to lock jaws, and either careen off in opposite directions, or one may grab another part of the opponent. If this happens the aggressor lets go and backs up. Alternatively, they may lock jaws, with each attempting to drag the other to the floor of the tank; usually after a few tugs by both, one animal struggles away and flees. Mouthfighting is the only type of fighting behavior in which an animal may be seriously injured. Three of the laboratory animals had their jaws dislocated in later experiments. Seitz (1948) observed a mouthfighting encounter between two very large *Tilapia* in which one fish killed the other within twenty minutes. Although he did not state the cause of death, this is a dramatic illustration of the possible result of mouthfighting. Oehlert (1958) suggests that, with many cichlids, lateral display, tailbeating, and mouthfighting do not do any damage. With *H. bimaculatus*, "the damaging thrust determines the issue within a few minutes," whereas, in *C. biocellatum* and *H. cyantognathus*, "rival display colors and behavior inhibit further fighting." She feels that other species, including *Tilapia*, are in between the two categories. Oehlert does not mention the size relationship between her animals. With respect to mouthfighting in *Tilapia*, her hypothesis is invalid only when the size difference is small.

The conclusion reached in these experiments is that every component of fighting behavior observed in group-raised males was performed by isolate males also.

The common sequences seen in every fight (e.g., carouseling and tailbeating on the descent) were similar to those performed by males that had been raised with other *Tilapia*.

Size of the Isolate Male in Relation to Coloration and Fighting

The standard lengths of males used in the eighteen successful encounters were from 3.7 to 5.6 cm. In the four encounters in which fighting did not occur, the size difference between animals was 0.8 cm or more. In three of the four, the reaction of the smaller animals to the larger courting males was to dash about the tank in Striped, Barred, or Hatched colorations for the duration of the observation hour. In the fourth encounter the larger animal swam slowly about the tank, ignoring the vigorously courting smaller fish. Thus the 0.8 cm difference between these animals appeared to prevent the smaller animals from threatening the larger, courting males, while conversely, the larger animal ignored the smaller, courting fish.

On calculating the difference in length between the eighteen pairs of animals which fought, in seventeen fights the larger animal won. In one encounter two threat movements were made by a fish 0.7 cm smaller than the other. Table 14 summarizes the data on size difference and fighting behavior of the 44 animals.

TABLE 14

Size Differences between Two Males in Relation to Total Time Spent Fighting, Fight Intensity, and Number of Bouts

Size difference cm S.L.[a]	No. of exper. in group	No. of encounters with no fight	Fight				Av. no. of bouts
			Av. duration in seconds	Intensity[b]			
				Strong	Medium	Weak	
0–0.2	5	0	1173 ± 407.4	4	1	0	17 ± 12
0.3–0.5	12	0	428 ± 76.2	4	5	3	4 ± 1.5
0.6–0.7	1	0	10 ± 0	0	0	1	1 ± 0
0.8–1.0	4	4	0	0	0	0	0

[a] These are not equal class intervals.

[b] In classifying the intensity of fights, those which included all components of fighting behavior were described as "strong," those which did not include mouthfighting but had all the other fighting behaviors were called "medium," and those which included only lateral display and tailbeating without contact were described as "weak."

The fights between animals differing from 0.3 to 0.5 cm were more variable in duration than those of the other groups. Generally the fights lasted longer and contained about four bouts. Animals differing in length between 0 and 0.2 cm fought continuously and strongly. In classifying the intensity of fights, those which included all components of fighting behavior are described as "strong," those which did not include mouthfighting but had all the other behaviors were called "medium," and those which included only lateral display and tailbeating without contact were considered "weak."

In all eighteen encounters the animals assumed the dark series of color patterns (11 in Dark 1, 23 in Dark 2, and 2 in Dark 3). "Strong" fights involved the Dark 2 and Dark 3 males, whereas in "medium" and "weak" fights the fish varied from

Dark 1 to Dark 2. Banded Black was not seen and Dark 2 was used for both fighting and courting. The two fish which became Dark 3 achieved this color phase for the first time after 20 minutes of fighting and faded to Dark 2 after the combat.

The average time taken to change from one color phase to the other was very short in isolate animals: 4 seconds from Striped to Neutral, 14.1 seconds to Dark 1, 36 seconds to Dark 2, and 41 seconds to Dark 3.[3]

Until the end of a fight, most of the individuals remained in the darkest color patterns they attained. All that were bitten and chased faded to Striped or Barred. When another fighting bout started, these fish darkened to their original fighting coloration. The average time taken to fade from Dark 2 to Striped was approximately the same as that recorded in table 4. From Dark 2 to Dark 1 it took 5.2 seconds; Dark 1 to Neutral, 3.7 seconds; and Neutral to Striped, 3.3 seconds. At the end of the fights all losers were lighter than the winners, and all but five were Neutral or Striped. All winners remained in the dark series of colors.

THE BEGINNING AND TERMINATION OF BOUTS

Fighting took place in a series of bouts. The fights classified as "weak" were short and consisted of only one bout. "Strong" fights consisted of more and longer bouts than "medium" fights, and generally lasted longer. In both "strong" and "medium" fights, the intervals between bouts lasted approximately as long as the bouts themselves. The onset of the next bout could not be predicted by the length of the rest interval. (See table 15.)

TABLE 15
TEMPORAL AND NUMERICAL RELATION BETWEEN BOUTS AND FIGHTS IN WHICH BOUTS OCCURRED

Fight intensity	No. of fights	Av. time of fight (secs.)	Av. time of actual fighting	Av. no. of bouts
Strong............	8	1293 ± 524.4	683.1 ± 137.2	16.1 ± 9.2
Medium...........	6	266.6 ± 42.1	135 ± 16.4	2.5 ± 1.43
Weak..............	4	31.7 ± 14.0	10.3 ± 14.0	1.0 ± 0.0

In an analysis of the onset of 122 bouts of fighting, 59 started after one of the animals began to court the other in the rest interval. Animals threatened the courting animal 47 times, and 12 times courting males threatened first. All courting animals threatened their opponents toward the end of the numerous bouts in strong fights when the courted animal began to darken in color. In the other 63 bouts, as one fish swam toward the other he was threatened and retaliated. Bouts were usually started when the fish met for the first time since the previous bout.

It was sometimes difficult to assign a cause to the termination of bouts. Of the 147 bouts recorded, 27 appeared to wane rather than to end abruptly. Perhaps fatigue of both fish was important.

When one or both animals swam away while carouseling, 63 of the bouts terminated. The nudging and biting must have had an influence in ending the bouts. The two animals parted by carouseling outward and wandering off. Often the upper fish could not get into the right position to tailbeat, and carouseling some-

[3] Averages taken on the second, third, and fourth color jump from Dark 2 to Dark 3.

times brought the two combatants to the water's surface. In either case the fighting was terminated.

The remaining 52 bouts were clearly caused by one animal's actions: 3 ended with mouthfighting, 5 with tailbeating, 8 with biting, and 37 with courting. When one fish suddenly stopped fighting and courted the other, the latter did not always rush away as in the bouts terminated by an aggressive act. In each of the 37 bouts the courted fish had started to fade from Dark 1 to Neutral.

Termination of Fighting Behavior of Isolate Males and Subsequent Activities and Color Changes

The fights were ended in two different ways. In eleven fights an aggressive act caused one individual to flee. One fight terminated in mouthfighting, two with butting, three with tailbeating, and five with biting. The winner chased the loser and after a few minutes courted him. The loser became Striped almost immediately. After being chased around the tank for a minute or two, he stopped and either took the raised position or hid in a corner. The chaser most frequently halted and began courting. In two of the fights the winner repeatedly nipped the motionless fish until the latter fled; he then resumed the chase and courted when his opponent hid in the corner in the Hatched coloration.

In the other seven fights one of the displaying animals began courting. Four of the courted animals continued to display laterally and finally fled. They turned Striped almost immediately. In three other fights the courted fish either nipped the substrate or swam away. When the tilting fish followed them, they gave lateral displays. Only one courting animal swam away from the threatening fish for a brief interval and then resumed courting.

Differences in Fighting Behavior of Previously Isolated versus Community-raised Males

In eight of this series of ten encounters, courting occurred after contact was made and in seven of these the isolate male courted first. Otherwise the behavior of males was the same in the two groups. Isolate males performed all the component fighting activities in the correct sequences, as did the community-raised males. Fighting began when the courted animal or the resident male threatened the other fish. Slight size differences between these small community-raised *Tilapia* and their isolate opponents determined fight intensity and duration as well as bout frequency. If the fish differed in size by more than 1 cm they did not fight. The termination of fights and of many of the bouts was courting, mouthfighting, butting, tailbeating, and biting. Some bouts ended during carouseling and may have been terminated because the proper orientation for tailbeating was not achieved. After fighting stopped, one animal courted the other or first chased and then courted him.

When the two males first began to threaten one another they became Dark 1. During the fight their color darkened to Dark 2 and Dark 3. The loser faded to Neutral, Barred, or Hatched and the winner became Dark 2 and courted. Two isolate animals darkened to Banded Black as did six community-raised males. No isolate male darkened before the community fish.

SUMMARY

In males isolated when they reached a length of 1–2 cm and tested when sexually mature, fighting and courting behavior was similar to that seen in animals raised together. The color changes were similar in the two groups, but the isolate males darkened more slowly than the community-raised males. Fighting between males took place in a number of bouts. Threat display led to a fight when one fish entered the other's territory and began to darken in color, or when one darkening male courted the other. Fights terminated when a male began to court a fading male, or as a result of mouthfighting, tailbeating, butting, or biting. In the latter cases, one animal fled and the other chased him and then courted.

DISCUSSION

COLORATION AND THE PREDICTION OF BEHAVIOR

Color change allows us to predict the future behavior of *Tilapia*. The dark series of color patterns are associated with courting and fighting, the striped series with fleeing. Dark 2 and Black occur most frequently when the males are courting and Dark 3 when they are fighting.

If *Tilapia* is photographed and the single frames are analyzed, in each frame are sets of cues that enable the investigator to determine not only what the animals will do next but what they have done in the past. If a fish is Dark 3 we can say that he has been threatening or fighting with his opponent and will continue this behavior for some seconds. If we then introduce our second source of evidence and say that the male is in midwater and has his body axis parallel to the substrate, all fins raised, an open mouth, and his branchiostegal rays spread, we can be more positive in our prediction. We may then introduce a third source of evidence and describe the other fish in the picture. The coloration of both individuals tells us that this frame was taken at some time between the middle and the end of a rather long fight. Since both fish are displaying laterally and are in midwater their next action will be to tailbeat. If, however, the second animal has all fins raised, has its mouth and branchiostegals closed, is oriented at an angle of 20° with its body axis pointed upward, and is Hatched, we can say that the animal has fled from its opponent and will remain in this position until our first animal courts or bites it. Comparing these two different frames we see that the position of the mouth and branchiostegal rays, the coloration, and the angle of orientation are diagnostic factors in contrasting the aggressive and fleeing tendencies of the animal.

When dealing with frames of single animals, the distinction between courting and fighting animals is based on fin position, angle of orientation, and coloration. A Black animal with all fins closed and oriented head downward is courting; a Dark 3 male with all fins raised and oriented parallel to the substrate has aggressive tendencies. If there are two males, the orientation of one with respect to the other is a diagnostic feature. If two Dark 3 males are shown mouthfighting the observer can determine the following: there had been a rather large fight of considerable duration; there have been many bouts of fighting; all behavioral components have been observed; the bout or fight is about to end; after the fight ends, one animal will flee in Striped, Barred, or Hatched, and the other will chase and then court in Black.

The prediction of behavior from color patterns depends on the extent to which the two may be correlated. Dark 1 indicates that a fish will soon court or fight, whereas the probabilities are high that Dark 3 fish will fight and Black fish will court. By adding certain motor patterns such as fin position to the observed color patterns, we can be more positive about our predictions of behavioral tendencies (e.g., Dark 1, raised fins = aggressive tendency; Dark 1, lowered fins = courting tendency).

FIN POSITION, ORIENTATION, AND COLORATION AS PREDICTORS OF BEHAVIOR

Fin position, orientation, and coloration are not unique to *T. mossambica* as predicators. Fin position as a diagnostic character for behavioral tendencies is particularly manageable. Morris (1958) used spine-raising in *Pygosteus* as a key factor in his discussion of motivational tendencies (raised spines = aggression, lowered spines = fright). He was then able to correlate a black coloration with raised spines (aggressive tendency) and a pale fish with lowered spines (fright tendency). Fin position has been used as a motivational parameter by many fish behaviorists. Several more conclusive studies of fin-spreading were conducted by Seitz (1948), Baerends and Baerends van Roon (1950), Wiepkema (1961), Barlow (1962a), and Wickler (1962). *Tilapia*'s lowered fins are typical of a courting animal. Raised fins correlate with aggression or fright. Barlow (1962a) has discussed conflicting fighting-fleeing tendencies in *Badis* by utilizing the raised fin-position and changes in coloration. The same characteristics apply to *Tilapia*, although they are used in a slightly different way. In both species a frightened animal has its branchiostegal rays and caudal fin closed, whereas a fighting animal's branchiostegals and caudal fin are spread. If a Dark 3 fighting *Tilapia* closes his caudal fin and branchiostegals he will become Striped and flee.

Variations in orientation in behavioral sequences are common among cichlids (Baerends and Baerends van Roon, 1950; Neil, pers. obs. of dwarf cichlid species). Wiepkema (1961) working on *Rhodeus,* Morris (1958) comparing *Pygosteus* and *Gasterosteus,* and other ethologists use it in support of their ideas on motivation. However, orientation is rather difficult to measure and therefore has limited application as a behavioral parameter. Our most accurate measurements of orientation were made when *Tilapia* is pointed away from (rising) or toward (tilting) the substrate. Rising is correlated with fear and tilting with courting. Measurements become subjective when *Tilapia* orients with respect to another fish. The orientation of a female following a courting male is similar to a fish chasing another fish. This parameter can be used in very few situations owing to numerous problems such as this.

Baerends *et al.* (1955) initiated the quantitative analysis of the color changes of fish. Certain color patterns in *Lebistes* were associated with specific behavioral acts, and thus gave clues to the motivational tendencies of the animal. Nelson (1962), working on male *Corynopoma riisei,* defined a sequence of courtship as a series of statistically dependent events bounded at either end by intervals separating temporally and qualitatively independent events. He found that the presence of a black belly spot was highly correlated with courtship sequences (per. com. Nelson). In this study he was able to intercorrelate three distinct variables to arrive at his conclusions. Barlow (1962a, 1962b, 1963) observed the color patterns of *Badis badis* in relation to specific behavioral components and was able to arrive at certain conclusions on motivational relationships. A fish which puts on more and more bars and finally becomes black shows an increasing tendency to court or fight. With *Badis*, as well as with many anabantid species (Forselius, 1957), courtship activities are highly correlated with aggressive tendencies. This may explain why fighting and courting *Tilapia* males are black.

Barlow has found that a *Badis badis* which sheds more and more disruptive coloration (eye slash, ocelli) and becomes pale is increasingly likely to be observed fleeing or motionless within or near cover. With *Tilapia*, fading from the courting and fighting colorations is correlated with the behaviors associated with a frightened fish (fleeing, rising, jerking). Concurrent with fading is the appearance of the Striped, Barred, and Hatched color patterns. Striped and Barred colorations are assumed in situations in which novel or threat stimuli are not repetitious. Animals placed in a small aquarium and repeatedly prodded with a net went into spasms of quivering and died. These fish invariably displayed the Hatched color pattern, which is the most extreme indication of fright. Hatched was observed also when fish were repeatedly bitten even after assuming the rising position, and when brooders had lost their young or were defending them. Thus *Tilapia*, unlike *Badis*, adds more disruptive color patterns as the number of novel or threat stimuli increase.

The Function of Courtship Coloration

Wherever communal spawning areas are found, in aquaria (Seitz, 1948; Baerends and Baerends van Roon, 1950), in natural waters (Lowe, 1959), or in fishponds (Chimits, 1955), the resident *T. mossambica* males with which the females spawn are large and Black. My field studies have indicated that the females outnumber the males two to one, whereas in the laboratory the ratio of male to female is the same. In the field, resident males are very easily spotted, whereas schooling fish are often difficult to locate. Chimits (1955) has noted the predominance of females in natural waters. Why can *Tilapia* afford a dramatic courtship coloration? The function of Black as a courtship coloration must outweigh its disadvantages in being conspicuous.

Although *Tilapia mossambica* has not been observed in the same spawning areas as other *Tilapia*, Lowe (1959) found that two closely related sympatric species of *Tilapia* had very different nuptial color patterns. *T. saka* was black with white median fin borders, whereas *T. squamipinnis* had a blue body and white head. These patterns were assumed only during courtship when the two species entered an area to spawn. Lowe feels that in this situation the male coloration is of importance in conspecific sexual recognition. If a male *T. mossambica* entered the same spawning grounds, his red fin borders and brown and white head would serve to label him as a distinct species.

In the laboratory, resident males of *T. mossambica* are Black. Males establishing territories may be Neutral or in the dark series of color patterns. Females will spawn with Dark 2 or Dark 3 males. There is no indication that Dark 3 males evoked a different reaction from the female, although the males did not remain Dark 3 long. Neutral and Dark 1 courting males were ignored. The white and red areas of the male courtship coloration do not stand out clearly in Dark 1 and in Neutral. Perhaps the female did not recognize these animals as courting males, since their body coloration was similar to her own. It was not possible to determine conclusively which of the elements of the male's dark coloration attracted the female, for she became very frightened when presented with models (Baerends and Baerends van Roon, 1950; Neil, per. exp.). In his comparative analysis of egg dummies of cichlid fishes, Wickler (1962) pointed out that the female *Tilapia* probably responds to the male's white genital papilla as a sign stimulus to approach. Thus we have one concrete and several suggestive pieces of evidence that the female recognizes male courtship coloration.

T. mossambica is among those species in which the male is more expendable than the female in that she carries the fertile eggs. A predator can spot a Black resident more readily than the female. If the resident sees this predator, perhaps his ability to fade rapidly will save him. In any event, the female would escape with her eggs. A replacement male is always available from the school.

The male of *T. macrocephala*, in contrast, broods the eggs (Aronson, 1949). Neither sex has a breeding coloration. Rather, the male and female remain a sandy color. Since they spawn on the substrate it would appear disadvantageous for the male to develop a brilliant coloration. Kühme (1961) found that the mouth-brooding male *Betta anabantoides* had a very inconspicuous courtship coloration with respect to its bubble-nest-building cousin *Betta splendens*. In *B. anabantoides* the male takes the eggs from the female in midwater, and the drab courtship coloration would protect this species as well. The male of several species of burrow-dwelling or nest-building fishes is responsible for rearing the eggs and young. A dark courtship coloration was reported by Barlow (1963) for *Badis*, by Morris (1954, 1958) for *Cottus* and *Pygosteus*, and by Tavolga (1956) for *Bathybobius*. These species often wait in or near their burrows or nests for females to appear. If a predator approaches, these males are in position to dive back into their nests, where the black color serves to conceal them.

Among the cichlids in which a male and female share the care of eggs and young, either parent can finish the job if one dies. During courtship of these species, both sexes are usually similar in coloration—either highly colored or dull. I have noticed this in several species of dwarf cichlids of the genus *Apistogramma*. It is reported also for *Cichlasoma bimaculatum, C. severum*, and *C. meeki* (Baerends and Baerends van Roon, 1950); for *Aequidens latifrons* (Breder, 1934); and for *Symphysodon discus* (Hildemann, 1959). Noble and Curtis (1939) found that the reddish-colored *Hemichromis bimaculatus* parents preferred red pots for spawning purposes. This is perhaps a protective device in that a parent could hide its pearly eggs by hovering over them.

In conclusion, the courtship coloration of *Tilapia* is thought to serve four major functions. It allows the female to recognize conspecific males and sympatric species to view them as unsuitable partners. It contrasts sufficiently with the substrate to allow localization at a distance. It can be changed rapidly to conceal a male from predators. It acts to divert the attention of a predator from females to the more expendable males.

THE NATURE OF DARK 3 COLORATION AND THE INTENSITY OF BEHAVIOR

Several discrepancies are associated with the correlations between Dark 3 and fighting, and between Dark 2 and Black and courting. Females will spawn with Dark 3 males, yet males changed from Dark 3 to Dark 2 to court. The courtship pattern of tailwagging was most commonly performed when the male was Dark 3. In their first encounters, isolate males fought for 20 minutes in Dark 2 and then became Dark 3. In all these encounters, fish were either courting or fighting in colorations contrary to those predicted. Further evidence may provide a solution to this problem.

A male performs the most active sequences of more than eight changes in courtship behavior per minute when he first sees a ripe female and when she first enters his nest. At these times, tailwaggings increase to 50 to 150 times the number performed during the rest of the time spent courting. In these most active sequences the male is most often Dark 3 and during the remainder of the time he is Dark 2 or Black. Dark 3 is therefore associated with high-intensity courting, as is tailwagging. After these active sequences the male changes to Black or Dark 2 and resumes courting. This indicates that these color patterns are correlated with lower thresholds of courting behavior. Wickler (1958) found that *Steatocranus* utilizes a checked color pattern both as protective and courtship coloration. He suggests that, since this cichlid holds its territory throughout the year, its coloration is correlated with lower thresholds of courtship. Although it has no defined breeding season, *Tilapia* will establish a territory for a week at a time. During the establishment of its territory it will court in Dark 2, which is not a truly protective coloration but is less noticeable than Dark 3. *Tilapia* can easily mask Dark 2 with stripes or bars, whereas it is more difficult to do this with Dark 3. Perhaps Dark 2 has a protective function superior to that of Dark 3 as well as a lower courtship threshold.

Dark 3 is assumed during high-intensity fighting (defined in table 14). Dark 2 and Black are used during the interval between and at the start of subsequent bouts. They are also used in weak- or medium-intensity fighting. In their first encounters with another male, young animals with nonexistent or half-completed pits tend to fight for some minutes in Dark 2. If they become Dark 3, mouthfighting results. Black residents with nests and females nearby will become Dark 3 more rapidly and only a few preliminary displays will be witnessed before mouthfighting terminates the encounter. Dark 3 represents a high threshold of fighting activity. Dark 2 is associated with medium or low aggressive activity.

Black has a slightly different function. Oehlert (1958), as a result of her studies on aggressive behavior of cichlids, feels that the more highly ritualized form of movement often has markedly lower thresholds than nonpredictable fighting behavior. This statement applies to pendeling in *Tilapia* because we can accurately predict the ensuing motor patterns. During intraterritorial fighting, however, there are several alternatives for what the animal will do next. Since pendeling represents a highly ritualized form of threat movement, its associated Black coloration is indicative of a low aggressive tendency.

On the basis of this evidence it seems that, in *Tilapia,* Dark 3 is associated with fighting and courting acts at the highest thresholds. Dark 2 is representative of middle and low courtship and aggressive tendencies. Black is typical of middle and low courtship tendencies and with low aggressive activity. The first aggressive encounters of territory-establishing males take much longer to reach high intensity than an encounter involving males with established nests and females close by. If Dark 3 fighting males begin to court, they become Dark 2 or Black. This indicates that there is a decrease in fighting tendency and a low level of courting activity.

THE PHYSIOLOGICAL BASIS AND RATE OF COLOR CHANGE

Several studies have been made on pigment change in *Tilapia*. The effects of hormones which are attributed to pigment cell multiplication and dispersion take longer than one minute to become visible (Pickford and Atz, 1957). The Black color pattern, which takes 24 hours to reach, is attributed to sexual hormones. Vertical cuts were made on the caudal fins of our *Tilapia*. Parker (1948) describes the resultant dark stripes as a neurohumoral effect which appears within an hour of incision. These streaks appeared in about 5 minutes and lasted three or four days. Since *Tilapia* take about 30 seconds to change from Hatched to Banded Black, far shorter than the described neurohumoral or hormonal effects, these patterns are thought to be under nervous control.

Wiepkema (1961) noted that the bitterling *Rhodeus* changed color gradually during fighting and became maximally colored within 5 to 10 minutes. He feels that these color changes are regulated by the autonomic nervous system—more specifically by the interaction of adrenaline and acetylcholine. It is very likely that the more rapid color changes of *Tilapia* are regulated by the autonomic nervous system.

Color patterns under nervous control may develop at different speeds. Morris (1958) notes that *Pygosteus* and *Gasterosteus* are slow to change color with respect to their behavior. *Badis* (Barlow, 1963), *Lebistes* (Baerends *et al.*, 1955), and *Tilapia* change color very rapidly. *Tilapia*'s color patterns are fewer and less distinct than those observed for *Badis* and *Lebistes*. Barlow described eleven distinct patterns and suggested that these could be broken down further. All three species, however, are able to change color before a change in behavior is observed.

SUMMARY

This paper contains a quantitative analysis of aspects of behavior related to the color changes of *Tilapia mossambica*. More than 1,000 animals were used in order to quantify the observational data. From the results, cues were established to predict *Tilapia*'s subsequent behavior from its color patterns or its subsequent color patterns from its behavior. Since fish longer than 20 cm are unable to lighten to their original youthful coloration, animals less than 15 cm long were used.

Tilapia mossambica, an extremely gregarious animal, seeks contact with members of its own species throughout life. At certain times only fish of a specific size range, coloration, or sex are accepted as companions. In the laboratory the young form tighter schools than the adults. This is due in part to the interference of courting and fighting males and brooding females. A minimum of ten young is needed to form a school in which all individuals are in Neutral coloration and are not fearful of the experimenter. Young less than 3 cm long are eaten by animals five times larger in the aquarium. Larger animals display to smaller ones if the size difference between them is less than 4 cm.

The first indications of a reproductively inclined male are Dark 1 coloration, courtship or threat behavior, and mouthdigging in a localized area. A territory is considered established if the male is Black, pendels with other resident males, and is in the territory approximately 90 per cent of the time. Preliminary observations of resident males indicate that there is a cyclic fluctuation between territory holding and schooling over a period of weeks. When there are few nest sites, only the largest males procure them.

Courting involves a definite sequence of events performed by both male and female. This is dependent on the correct response of the female. If she is not receptive, the male repeats a previous step in the courtship sequence.

When the male first meets the female and when she first enters his nest, the intensity of his courting increases to eight changes in courtship behavior per minute. On both of these occasions the highest number of tailwags per minute were observed. At these times the female became motionless and fled when the male tailwagged. Perhaps tailwagging is essential in order to prevent "freezing" on the part of the female. The correct responses by the female become more frequent as time progresses and the courtship sequence becomes briefer. The 6 cm male tilts, twitch-leads, or leads when the female is in midwater or on the substrate some distance from the nest. These activities are replaced by rolling when the female is on the substrate within 20 cm of the nest or in midwater within 15 cm above the nest. The male constantly attempts to show his lateral profile to the female. Rolling accomplishes this aim when the female is too near the nest for the tilting position to produce the same result.

In the laboratory, spawning occurred at a minimum of 1½ hours after the first visit of the female to the nest. The female spawns with more than one male, but prefers the territory in which she digs for the longest period of time and helps defend. A ripe female will respond in the correct manner to Dark 2, Dark 3, and Black courting males.

The insertion of females into a tank results in an increase in Black males. New males or females produce an immediate increase in courting by resident males. Resident males tilt to new males more often than to familiar females. After the first few hours, the courtship of new males decreases rapidly as the strangers begin to darken and threaten the residents. From this time the females are courted more often.

Brooding females in the community tanks become Striped at about the time the young hatch in their mouths. The females become Hatched when the eject their young for the first time. When all the young are eaten by other fish the mother guards and "calls" air bubbles for a few days. Isolated brooders are Neutral or Striped and have dark mouths and eyes. When the young begin to wander and are not snapped up by their mother in the face of danger, the females become light around the mouth and eyes.

Fighting between males takes place in a number of bouts. Threat display leading to a fight begins when one dark fish enters the other's territory or when one male courts the other. Fights and bouts within fights are terminated when one male fades and the other begins to court, or as a result of mouthfighting, tailbeating, butting, or biting. In the latter cases, one animal flees and the other chases him and then courts.

The intensity, number of bouts, and length of a fight between young males depend in part on the size difference between animals. Animals between 3.7 and 5.6 cm in length do not fight if the difference in their length is greater than 0.8 cm. The larger animal courts the smaller.

Males isolated one week after hatching and tested when sexually mature displayed the same fighting and courting behavior as that seen in animals raised together. The color changes were similar in the two groups, but isolate males darken more slowly than community-raised males. The number of courtship activities per minute was higher for the isolate males.

Reproductive color patterns and their relation to the behavior of *T. mossambica* are discussed throughout the paper. A male becomes Dark 1 immediately after he begins to court or fight. Fairly high and consistent correlations have been established for Black as a courting color and Dark 3 as an intraterritorial fighting color. A male becomes lighter when he is chased or when he leaves his territory, and darkens when he chases other males or returns to his territory. Males darken when they fight and court. Black fish become Dark 3 when they fight and Dark 3 courting animals, if unable to change to Black, become Dark 2. Pendeling is threat behavior between Black resident males. This behavior has become ritualized, since it follows the same temporal pattern each time and no contact is made between males. Resident males who chase may be either Black or Dark 3.

All color changes with the exception of Black are very rapid. Since it requires over 24 hours to become Black, the animals court in Banded Black. The difference between the two is detected by other males, for the nests of Banded Black males are attacked and those of Black males are seldom approached unless there is a female present. The number of bouts, and the presence or absence of mouthfighting, butting, biting, and tailbeating determine the intensity of fighting behavior. The males become Dark 3 during medium- to high-intensity fights.

LITERATURE CITED

ARONSON, LESTER R.
1949. An analysis of reproductive behavior in the mouthbreeding cichlid fish, *Tilapia macrocephala* (Bleeker). Zoologica, 34:133-158.
1951. Factors influencing the spawning frequency in the female cichlid fish, *Tilapia macrocephala*. Amer. Mus. Nov. 1484:1-26.

ATZ, J. W.
1954. The peregrinating tilapia. Animal Kingdom, 57:148-155.

BAERENDS, G. P., and J. M. BAERENDS VAN ROON
1950. An introduction to the study of the ethology of cichlid fishes. Behaviour, Suppl., 1: 1-242.

BAERENDS, G. P., R. BROUWER, and H. T. J. WATERBOLK
1955. Ethological studies on *Lebistes reticulatus*. I. An analysis of the male courtship pattern. Behaviour, 8:17-334.

BARLOW, GEORGE W.
1962a. Ethology of the Asian teleost *Badis badis*. III. Aggressive behavior. Zeit. für Tierpsch., 19:29-55.
1962b. Ethology of the Asian teleost *Badis badis*. IV. Sexual behavior. Copeia, 2:345-359.
1963. Ethology of the Asian teleost *Badis badis*. II. Motivation and signal value of the color patterns. Animal Behavior, 11:97-105.

BRAWN, V. M.
1961. Aggressive behavior in the cod (*Gadus callarias* L.). Behaviour, 18:107-147.

BREDER, C. M.
1934. An experimental study of the reproductive habits and life history of the cichlid fish *Aequidens latifrons* (Steindachner). Zoologica, 28:1-42.
1951. Studies of the structure of a fish school. Am. Mus. Nat. Hist. Bull., 98:7-27.

BROWN, M. E. (ed.)
1957. Physiology of fishes. New York: Academy Press. I, 246-277; II, 229-304, 367-401.

CHIMITS, P.
1955. *Tilapia* and its culture. E. A. O. Fish. Bull. 8:1-35.

COPLEY, H.
1958. Common freshwater fishes of east Africa. London: H. F. and G. Witherby, Ltd.

FORSELIUS, STEN
1957. Studies of anabantid fishes. Zool. Bidrag. fran Uppsala, 32:93-599.

HILDEMANN, W. H.
1959. A cichlid fish, *Symphosodon discus*, with unique nurture habits. Am. Nat., 93 (868): 27-34.

HINDE, R. A.
1959. Unitary drives. Animal Behavior, 7:130-142.

KEENLEYSIDE, M. H. A.
1955. Some aspects of schooling behavior of fish. Behaviour, 8:183-249.

KÜHME, W.
1961. Verhaltensstudien am maulbrütenden (*Betta anabantoides* Bleeker) und am nestbauenden Kampffisch (*B. splendens* Regen). Zeit. für Tierpsch., 18:33-55.
1962. Das Schwarmverhalten elterngeführter jungen Cichliden (Pisces). Ibid., 19:513-538.

LOWE, R. H.
1959. Breeding behavior patterns and ecological differences between *Tilapia* species and their significance for evolution within the genus *Tilapia* (Pisces; Cichlidae). Proc. Zool. Soc. London, 132:1-31.

MARUYAMA, T.
1948. An observation on *Tilapia mossambica* in pond referring to diurnal movement with temperature change. Bull. Freshw. Fish. Res. Lab. Tokyo, 8:25-32.

MORRIS, D.
1954. Reproductive behavior of the river bullhead, *Cottis gobio* with special reference to fanning activity. Behaviour, 7:1-32.
1958. The reproductive behavior of the ten-spined stickleback (*Pygosteus pugitius*). Ibid., Suppl., 6:1-154.

NELSON, KEITH
1962. Abstract 223. Am. Zool., 2(3).

NOBLE, G. K., and B. CURTIS
1939. The social behavior of the jewel fish, *Hemichromis bimaculatus*. Am. Mus. Nat. Hist. Bull., 76:1-46.

OEHLERT, BEATRICE
　1958. Kampf und Paarbildung einger Cichliden. Zeit. für Tierpsch., 15:141–174.
PARKER, G. H.
　1948. Animal color changes and their neurohumors. Cambridge: Univ. Press.
PICKFORD, G. E., and J. W. ATZ
　1957. The physiology of the pituitary gland of fishes. New York Zoölogical Society.
RUWET, JEAN CLAUDE
　1963. Observations sur le comportement sexuel de *Tilapia macrochir* Boulanger (Pisces: Cichlidae) au lac de retinue de la Lufira (Katanga). Behaviour, 20:243–250.
SEITZ, A.
　1940. Die Paarbildung bei einigen Cichliden. 1. Die Paarbildung bei *Astatotilapia strigigena* (Pfeffer). Zeit. für Tierpsch., 4:40–84.
　1948. Vergleichende Verhaltensforschung an Buntbarschen. *Ibid.*, 6:202–325.
SHAW, EVELYN
　1960. The development of schooling behavior in fishes. Physiol. Zool., 33:79–86.
TAVOLGA, W. N.
　1956. Visual, chemical, and sound stimuli as cues in the sex discriminatory behavior of the gobiid fish, *Bathygobius soporator*. Zoologica (N. Y.), 41:49–65.
TINBERGEN, N., and J. J. A. VAN IERSAL
　1947. "Displacement reactions" in the three-spined stickleback. Behaviour, 1:56–63.
WICKLER, WOLFGANG
　1958. Vergleichende Verhaltensstudien an Grundfischen. II. Die Spezialisierung des *Steatocranus*. Zeit. für Tierpsch., 15:427–446.
　1960. Belegexemplare zu Ethogrammen. *Ibid.*, 17:141–142.
　1962. Ei-Attrappen und Maulbruten bei afrikanischen Cichliden. *Ibid.*, 19:129–164.
WIEPKEMA, P. R.
　1961. An ethological analysis of the reproductive behavior of the bitterling (*Rhodeus amarus* Bloch). Arch. Neerland. Zool., 15:103–199.

PLATES

PLATE 1

Color pattern of the reproductive male *Tilapia mossambica*.

A. Neutral

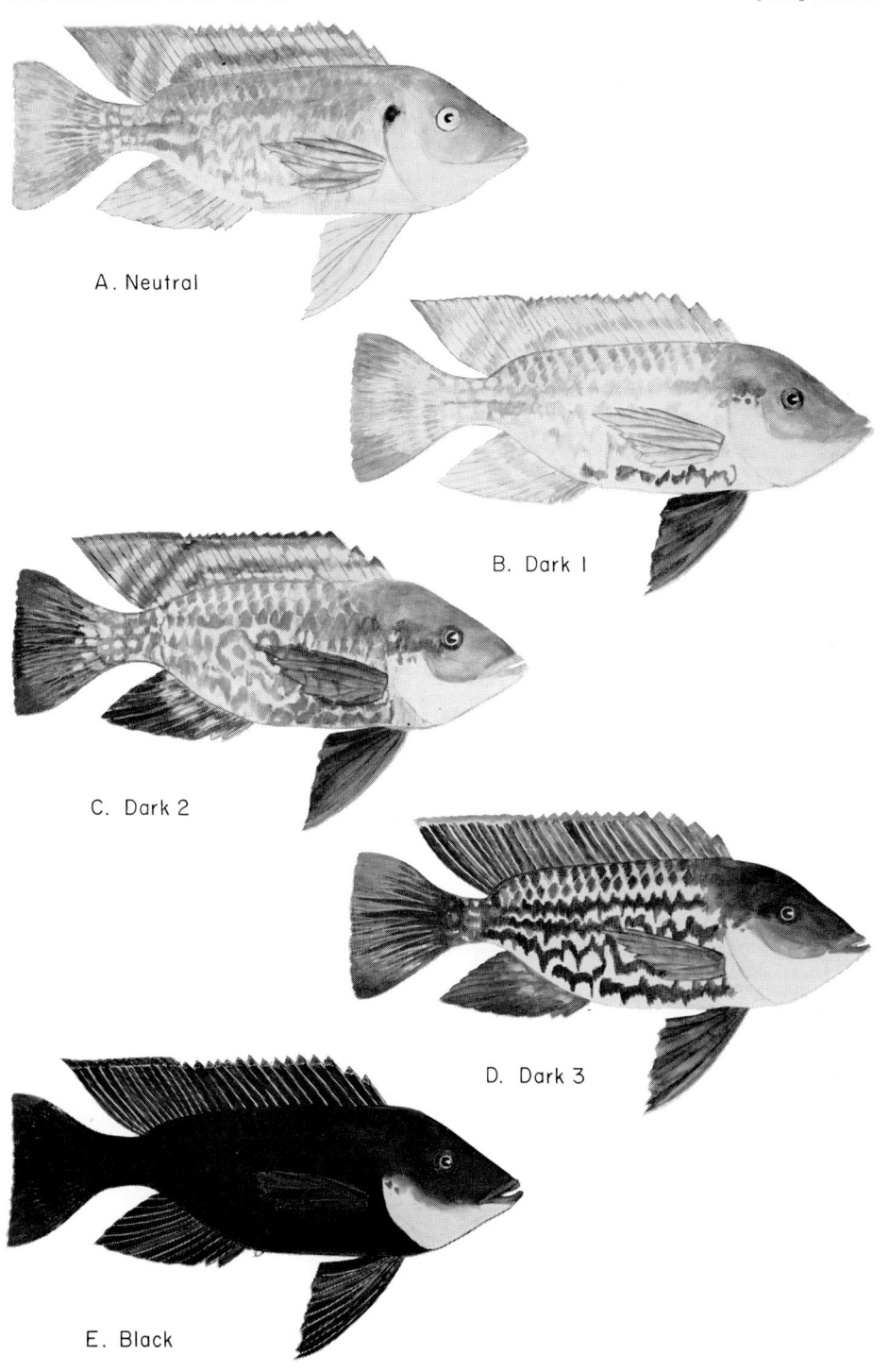

B. Dark 1

C. Dark 2

D. Dark 3

E. Black

PLATE 2

Nonreproductive color patterns of *Tilapia mossambica*.

A. Neutral

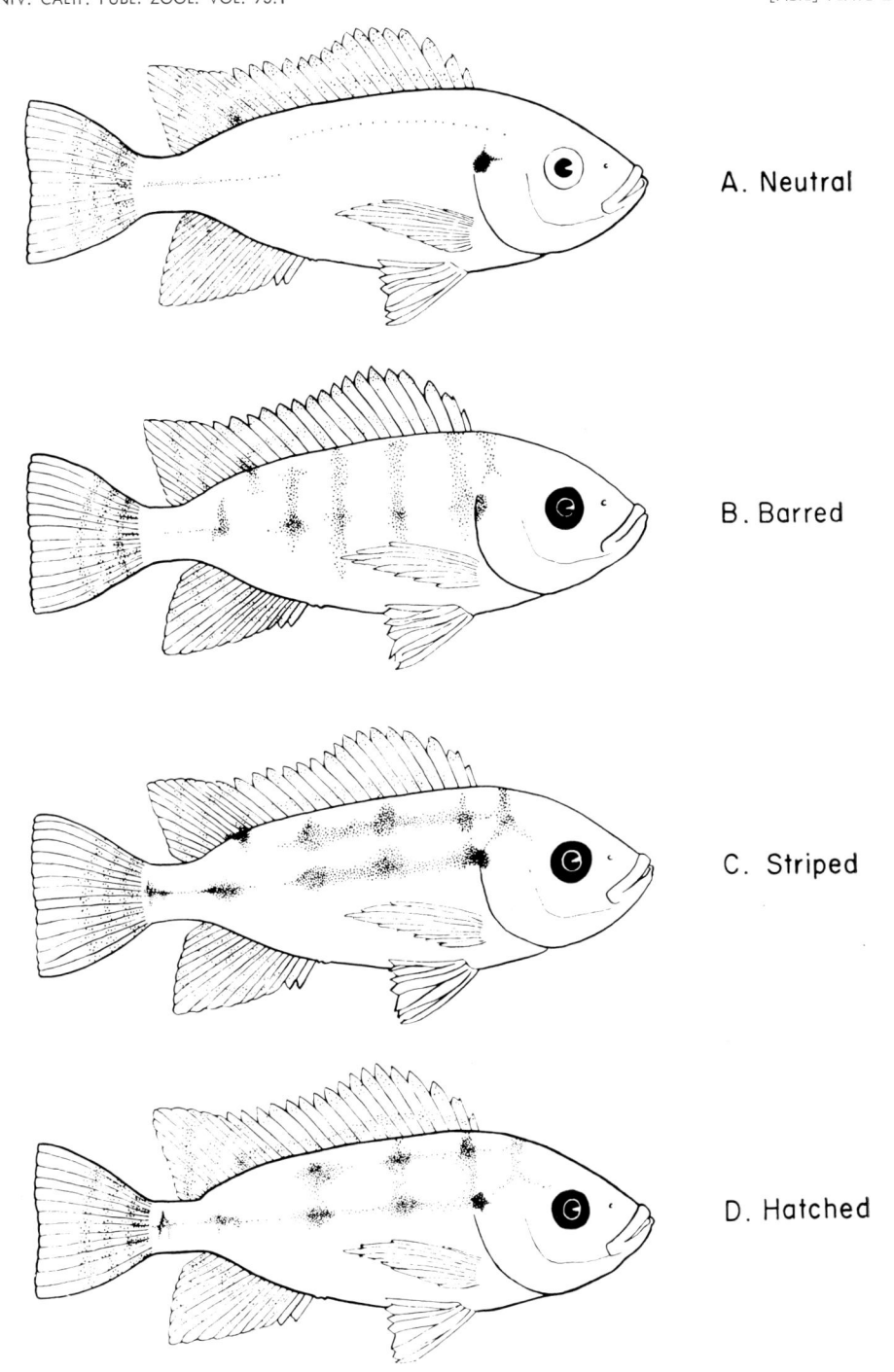

B. Barred

C. Striped

D. Hatched

PLATE 3

Color patterns of *Tilapia mossambica*: a, Neutral male; b, Dark 1 male leading Neutral female; c, Dark 2 male tilting; d, Dark 3 male, lateral display.

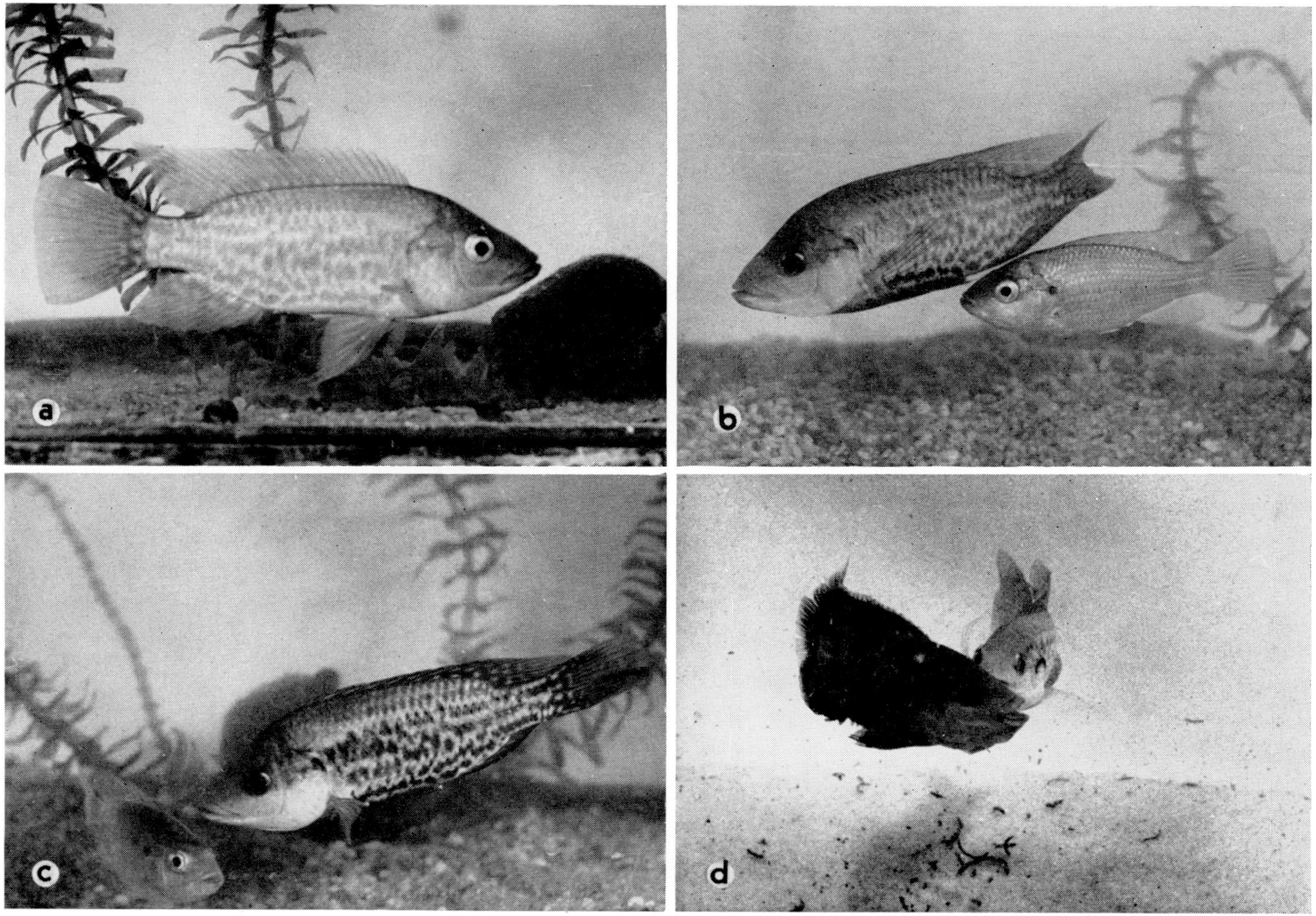
a
b
c
d

PLATE 4

Color patterns of *Tilapia mossambica*: *a*, Black male, lateral display, and Striped male rising; *b*, Black male rolling, Neutral female mouthdigging; *c*, Neutral female with fins raised, Dark 3 male tailwagging; *d*, Neutral young with "*Tilapia* mark."

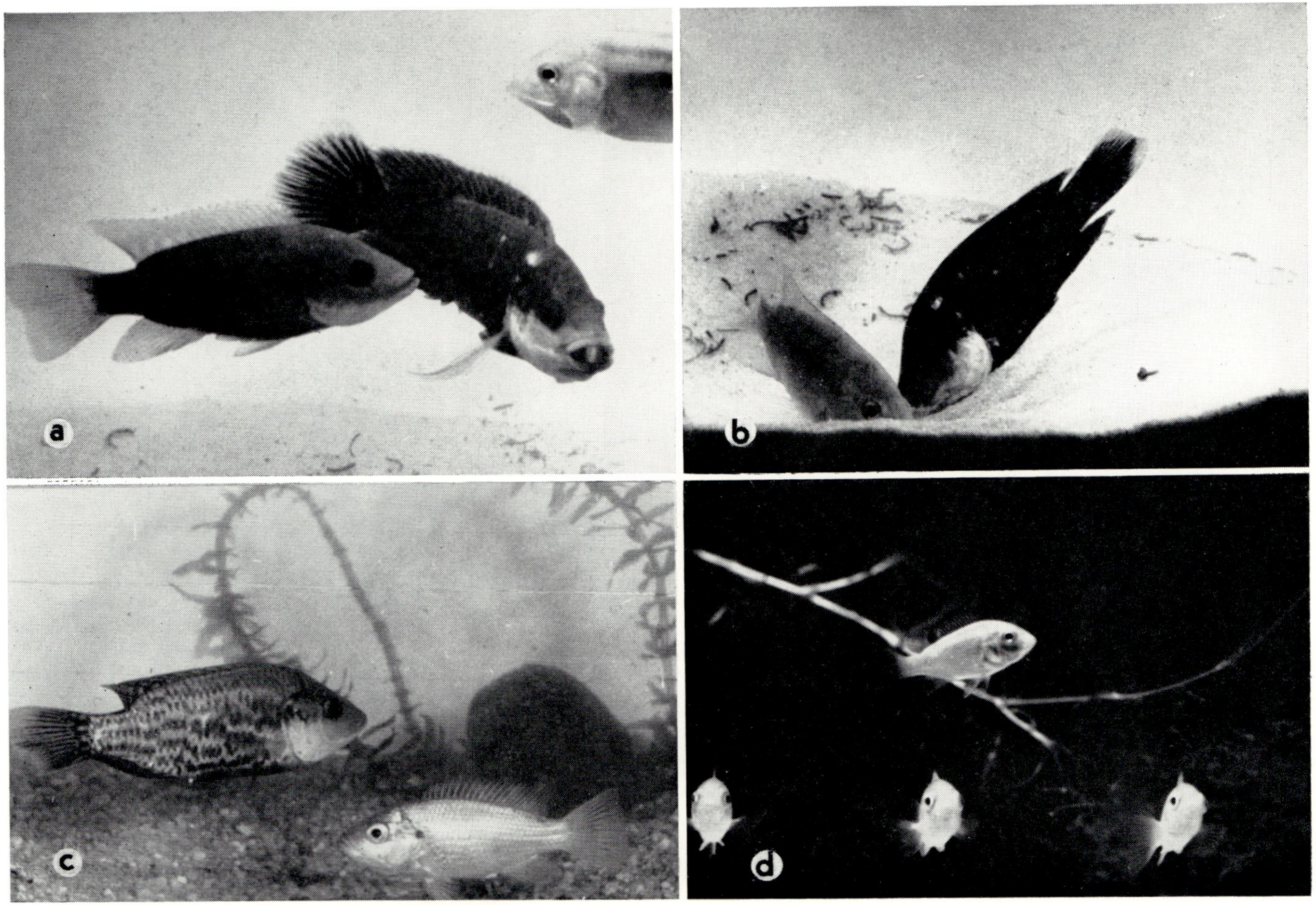
a
b
c
d

PLATE 5

Color patterns of *Tilapia mossambica: a*, Barred female fleeing; *b*, Striped female "calling her young"; *c*, Hatched female butting; *d*, Neutral females brooding.

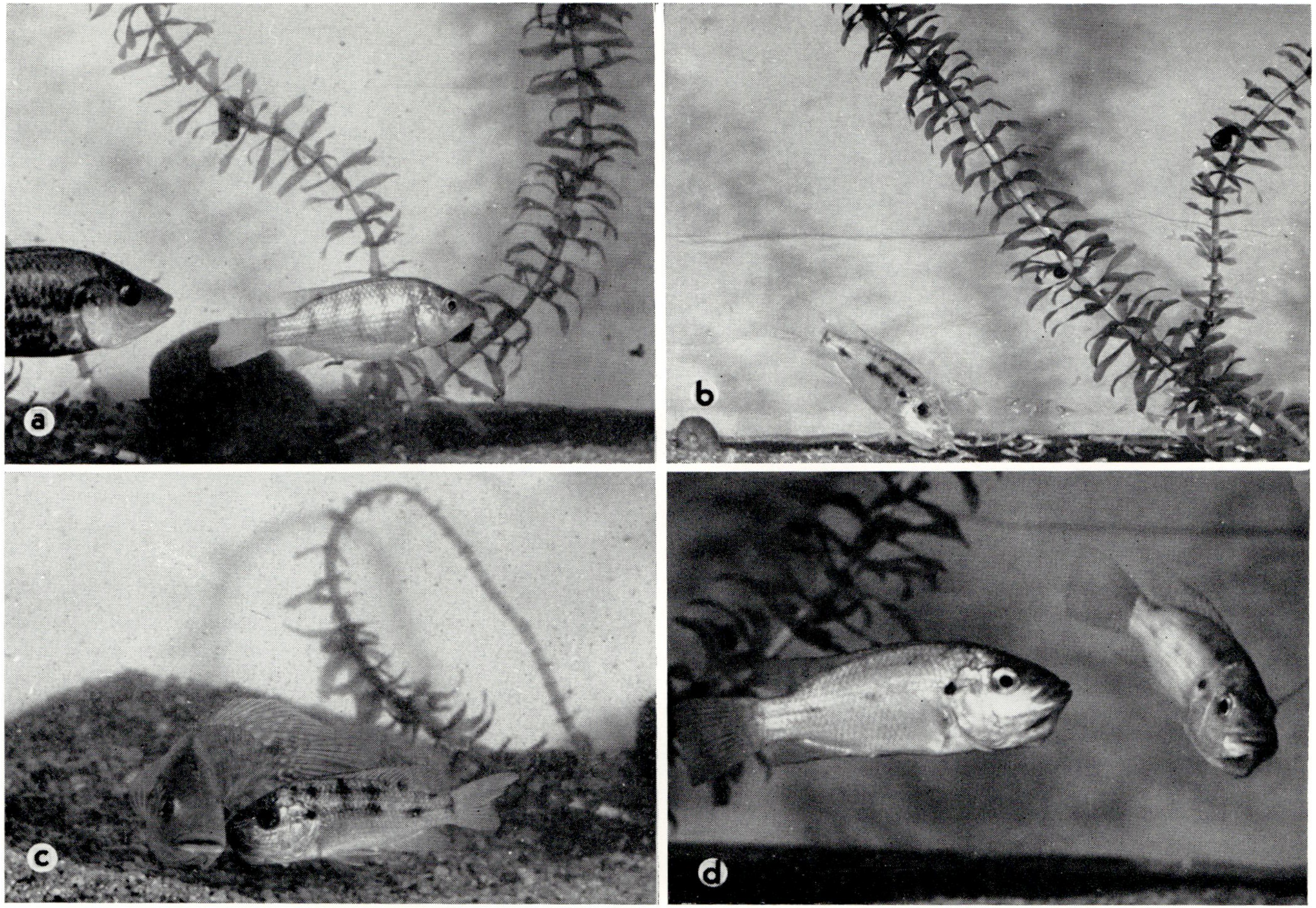
a
b
c
d